Risk
Assessment
with
Time to Event
Models

T0179095

Environmental and Ecological Risk Assessment Series

Series Editor
Michael C. Newman
College of William and Mary
Virginia Institute of Marine Science
Gloucester Point, Virginia

Published Titles

Coastal and Estuarine Risk Assessment
Edited by
Michael C. Newman, Morris H. Roberts, Jr., and Robert C. Hale

Risk Assessment with Time to Event Models
Edited by
Mark Crane, Michael C. Newman, Peter F. Chapman, and John Fenlon

Species Sensitivity Distributions in Ecotoxicology
Edited by
Leo Posthuma, Glenn W. Suter II, and Theo P. Traas

Risk
Assessment
with
Time to Event
Models

Edited by
Mark Crane
Michael C. Newman
Peter F. Chapman
John Fenlon

CRC Press
Taylor & Francis Group
Boca Raton London New York

CRC Press is an imprint of the
Taylor & Francis Group, an **informa** business

CRC Press LLC
Taylor & Francis Group
6000 Broken Sound Parkway NW, Suite 300
Boca Raton, FL 33487-2742

First issued in paperback 2020

ISBN 13: 978-0-367-57871-8 (pbk)
ISBN 13: 978-1-56670-582-0 (hbk)

Visit the Taylor & Francis Web site at
http://www.taylorandfrancis.com

and the CRC Press Web site at
http://www.crcpress.com

Library of Congress Cataloging-in-Publication Data

Risk assessment with time to event models / editors, Mark Crane ... [et al.].
    p. cm.-- (Environmental and ecological risk assessment)
Includes bibliographical references and index.
ISBN 1-56670-582-7
    1. Environmental risk assessment--Methodology. I. Crane, Mark. II. Series.
GE145 .R583 2001
363.7'02—dc21
                                                                        2001038737
                                                                        CIP

Library of Congress Card Number 2001038737

Preface

Polluting substances poison plants and animals over different spatial and temporal scales. The disciplines of environmental toxicology and environmental chemistry recognize this, and practitioners in these fields investigate toxic chemicals and their effects from the molecular to the ecosystem scale, over periods of time ranging from seconds to decades.

However, it remains true that most prospective risk assessments designed to test the safety of chemicals *before* they are released into the environment depend upon results from toxicity tests with single species performed in laboratories over periods ranging from a few hours to a few days. The results from these tests are reported as lethal concentrations (LCx), effective concentrations (ECx) or no observed effect concentrations (NOECs) at the end of the test periods. Hence, the literature is filled with examples of 48-h EC50, 96-h LC50, and 21-d NOEC values.

Unfortunately, environmental pollution does not operate on a 24-hour cycle. Many pollution incidents occur over periods of time ranging from minutes to hours. A short release of toxic effluent into a river may travel down that river as a discrete "plug." Tidal influences may carry pollution to and fro over the same point in an estuary within the space of a few hours. A spillage at sea may be diluted to low levels in a matter of minutes. How can an environmental regulator use information on 48-hour toxicity to predict the impacts of a few minutes of pollution in cases like these? At the other extreme, what is the relevance of 96-hour toxicity data for organisms that may have been exposed to a pollutant for 6 months below a continuous effluent discharge, or in soil contaminated with a persistent chemical?

The key to answering this type of question lies in the time course of toxic effect. Laboratory scientists do not set up their toxicity experiments and then go for a 4-day coffee break before returning to examine the results. Regular checks are made on test systems to ensure that an experiment is proceeding smoothly, and, in most laboratories, these checks are recorded as numbers of organisms dead, or numbers of offspring produced at different time points during the experiment.

This book explains how data collected in this way can be used to improve risk assessment, by using time to event models that are common in other disciplines but have only occasionally been used in environmental toxicology and chemistry. The idea for producing a book emerged after a meeting held in April 1998 at Horticulture Research International, Wellesbourne, Warwickshire, UK. This meeting, "Improving Risk Assessment with Time to Event Models" was sponsored by the Society for Environmental Toxicology and Chemistry, and brought together most of the authors of chapters in this book, plus experts from industry, academia and government regulatory agencies, with backgrounds in ecotoxicology or statistics.

Chapter 1 of this book provides a general background to time to event analysis and explains why the approach is important for risk assessment. Chapter 2 has some simple examples of the use of time to event analysis using the types of results that are common in laboratory ecotoxicity tests used for risk assessment. Chapter 3 extends this theme to populations. These chapters show how a careful examination of the time course of toxicity during an experiment can help risk assessors tease out more useful information than is present in a simple summary statistic such as a 48-h LC50.

In Chapter 4, the focus changes to use of time to event approaches to extrapolate to long term toxicity, and, in Chapter 5, the usefulness of time to event models for increasing the precision of toxicity summaries is explained. Chapter 6 presents a pragmatic approach for incorporating time of exposure into risk assessments for discharges from oil and gas platforms.

Time to event approaches have been used for years in other disciplines that share the types of problems addressed by environmental risk assessors. The way these approaches aid agricultural scientists (Chapter 7), ecologists (Chapter 8) and engineers (Chapter 9), should assist environmental toxicologists and chemists in recognizing the value of time to event techniques, and the wider arena in which they are used.

The specific use of time to event analyses in environmental risk assessment is addressed in Chapter 10, which is a summary of the Wellsbourne workshop. Overall conclusions and recommendations are provided in Chapter 11.

We hope that this book contains sufficient evidence to persuade workers in the field of environmental risk assessment to consider more explicitly the value of time when analyzing and reporting toxicity data. The statistics of time to event analysis need not be complex, while the understanding gained can be considerable.

Mark Crane
Michael C. Newman
Peter F. Chapman
John Fenlon
June, 2001

Acknowledgments

The editors thank the authors and participants at the Wellesbourne workshop for donating their time and expertise to this project. We thank Randi Gonzalez, April Heaton, Sylvia Wood and David Packer of CRC Press/Lewis Publishers for support during editing and publication. We also thank the referees for assuring the quality of the chapters presented here.

Referees

John Bailer
Miami University
Miami, Ohio

Jacques Bedaux
Vrije Universiteit
Amsterdam, The Netherlands

Peter Chapman
Syngenta
Bracknell, UK

Bob Clarke
Centre for Coastal and Marine
  Sciences
Plymouth, UK

Mark Crane
Royal Holloway
University of London
Egham, UK

Philip M. Dixon
Iowa State University
Ames, Iowa

John Fenlon
Horticulture Research International
Wellesbourne, UK

Ann Gould
Melrose, UK

Alan Kimber
Reading University
Reading, UK

Herbert Koepp
Federal Biological Research Centre
  for Agriculture and Forestry
Braunschweig, Germany

Eddie McIndoe
Syngenta
Bracknell, UK

Michael C. Newman
College of William and Mary
Williamsburg, Virginia

Jim Oris
Miami University
Miami, Ohio

Simon Pack
Procter and Gamble
Egham, UK

Joe Perry
Institute of Arable and Crop
  Research
Herpenden, UK

Keith Solomon
University of Guelph
Guelph, Ontario, Canada

John Thain
Centre for Environment, Fisheries
and Aquaculture Science
Burnham, UK

Paul J. Van den Brink
DLO Winand Staring Centre for
Integrated Land, Soil and Water
Research
Wageningen, The Netherlands

Paul Whitehouse
WRC-NSF
Medmenham, UK

John Wiles
DuPont
Stevenage, UK

Editors

Mark Crane is a senior lecturer in environmental biology in the School of Biological Sciences, Royal Holloway, University of London. He received a B.Sc. in ecology from the University of East Anglia, and a Ph.D. in ecotoxicology from the University of Reading. He worked for the Water Research Centre in the UK as an ecotoxicologist before joining the faculty at Royal Holloway in 1994 as a lecturer. He became the Director of Graduate Study in 2000.

Dr. Crane's research interests include probabilistic approaches to environmental risk assessment, invertebrate endocrine disruption, biochemical biomarkers of pesticide exposure, development of *in situ* water and sediment bioassays, and river quality monitoring in Europe and developing countries. He has published more than 50 articles on these topics, and co-edited two books, *Endocrine Disruption in Invertebrates: Endocrinology, Testing and Assessment* (1999), and *Forecasting the Environmental Fate and Effects of Chemicals* (2001).

Michael C. Newman is a professor of marine science and the Dean of Graduate Studies at the College of William and Mary's Virginia Institute of Marine Science/School of Marine Science (VIMS/SMS). After receiving B.A. and M.S. (Zoology with marine emphasis) degrees from the University of Connecticut, he earned M.S. and Ph.D. degrees in Environmental Sciences from Rutgers University. He joined the faculty at the University of Georgia's Savannah River Ecology Laboratory (SREL) in 1983, becoming an SREL group head in 1996. Leaving SREL to join the VIMS/SMS faculty in 1998, he took up his present position in 1999.

Dr. Newman's research interests include quantitative methods for ecological risk assessment and ecoepidemiology, population responses to toxicant exposure including genetic responses, QSAR-like models for metals, bioaccumulation and toxicokinetic models for metals and radionuclides, toxicity models including time to death models, and environmental statistics. He has published more than 85 articles on these topics. He has authored three books: *Quantitative Methods in Aquatic Ecotoxicology* (1995), *Fundamentals of Ecotoxicology* (1998), and *Population Ecotoxicology* (2001), and co-edited three books: *Metal Ecotoxicology: Concepts and Applications* (1991), *Ecotoxicology: A Hierarchical Treatment* (1996), and *Risk Assessment: Logic and Measurement* (1998).

Peter F. Chapman is based at Jealott's Hill International Research Centre in Bracknell, UK. After graduating in mathematics from Manchester University, he received a Ph.D. in applied statistics from the University of Exeter in 1977. He joined the Grassland Research Institute in 1977 and spent more than 5 years working on statistical problems in field crop experimentation before joining ICI Plant Protection Division (later ICI Agrochemicals, then Zeneca Agrochemicals, and now Syngenta) in 1982. He was appointed head of statistics for ICI Agrochemicals in 1987. At ICI and Zeneca, he worked on a wide variety of statistical problems, initially concentrating on crop protection, but later specializing in environmental safety. With the formation of Syngenta in November 2000, he was appointed head of Environmental Statistics.

Dr. Chapman's current interests relate to the application of statistical methods in the following areas: gene expression; laboratory and field ecotoxicological testing; multi-species field experiments; fate of chemicals in soils, crops and the environment generally; and environmental and human health risk assessment. He has been active in developing and promoting the use of statistical methods in environmental toxicology. He is an active member of the Society for Environmental Chemistry and Toxicology, and has organized statistical meetings and workshops on behalf of both SETAC and the OECD. He is an active member of the Royal Statistical Society and is currently chairman of the Reading, UK group of the RSS.

John S. Fenlon is head of the Biometrics Department at Horticulture Research International (HRI) in Wellesbourne, UK. After graduating in mathematics he received an M.Sc. in mathematical statistics from the University of Manchester in 1969. His career has covered posts in agriculture, horticulture and environment. In 1983, he was appointed head of statistics at the Glasshouse Crops Research Institute at Littlehampton, UK, and 5 years later became head of biometrics of the expanded organization that became known as HRI.

Mr. Fenlon's research interests relate mainly to the design and analysis of bioassay experiments in the field of biological control, particularly with regard to time to response. Areas of current activity include the analysis of quantal response data in invertebrate pathology, modeling spatiotemporal effects in predator–prey systems and the development of probabilistic methods for ecological risk analysis. A particular interest is the development of stochastic models to explain the variability of predator–prey and pathogen systems. He has published more than 50 scientific papers in many areas of applied statistics, and has been an active member of the Royal Statistical Society.

Contributors

Denny R. Buckler
U.S. Geological Survey
Columbia Environmental
  Research Center
Columbia, Missouri

Peter F. Chapman
Syngenta
Jealott's Hill Research Station
Bracknell
Berkshire, UK

Mark Crane
School of Biological Sciences
Royal Holloway
University of London
Egham, Surrey, UK

Philip M. Dixon
Department of Statistics
Iowa State University
Ames, Iowa

Mark R. Ellersieck
University of Missouri
Agricultural Experiment Station
Columbia, Missouri

John Fenlon
Horticulture Research International
Wellesbourne
Warwick, UK

Albania Grosso
National Groundwater and
  Contaminated Land Centre
  Environment Agency
Olton Court
Olton
Solihull, UK

Chris C. Karman
TNO Institute of
  Environmental Sciences
Energy Research and Process
  Innovation
Den Helder
The Netherlands

Alan Kimber
Department of Applied Statistics
University of Reading
Reading, UK

Gary F. Krause
University of Missouri
Agricultural Experiment Station
Columbia, Missouri

Gunhee Lee
University of Missouri
Agricultural Experiment Station
Columbia, Missouri

Bryan F.J. Manly
Western EcoSystem Technology Inc.
Laramie, Wyoming

Foster L. Mayer
U.S. Environmental
  Protection Agency
Gulf Ecology Division
  (NHEERL-ORD)
Gulf Breeze, Florida

John T. McCloskey
Savannah River Ecology Laboratory
University of Georgia
Aiken, South Carolina

Michael C. Newman
Department of
  Environmental Sciences
The College of William and Mary
Virginia Institute of Marine
  Science/School of Marine Science
Gloucester Point, Virginia

Tim Sparks
Centre for Ecology and Hydrology
Monks Wood
Abbots Ripton
Huntingdon
Cambridgeshire, UK

Kai Sun
University of Missouri
Agricultural Experiment Station
Columbia, Missouri

Contents

chapter 1

Introduction to time to event methods

Michael C. Newman and Mark Crane

Contents

1.1 Introduction

Conventional effects metrics, e.g., LC/EC50 and NOEC, focus on exposure intensity while holding duration of exposure constant; exposure intensity is estimated precisely but exposure duration is grossly fixed as acute or chronic. Although useful in assigning relative toxicities to various chemicals, or to the same chemical under different conditions, such information is temporally compromised if accurate prediction of exposure effect is a requirement. This, of course, is a major task in ecological risk assessment and, therefore, better inclusion of exposure duration would improve many risk assessments. This chapter briefly introduces conventional time to event methods that allow simultaneous inclusion of exposure intensity and duration in effects predictions.

1.2 What are time to event methods?

> *All substances are poisons; there are none which is not a poison. The right dose differentiates a poison from a remedy.*
>
> —Paracelsus

1-56670-582-7/02/$0.00+$1.50
© 2002 by CRC Press LLC

Rendered to "The dose makes the poison, " Paracelsus' admonition is a helpful reminder that both quality and quantity of an administered substance determines whether a poisoning will occur. Unfortunately, some toxicologists forget the context of this statement, remaining convinced that dose or concentration quantification is the most crucial task in determining exposure consequences. This preoccupation extends to toxicologists interested in consequences to ecological receptors of environmental exposures.

Paracelsus was concerned with ingested doses and that is the appropriate context for applying his maxim. However, many toxicologists have become preoccupied with doses and concentrations to the neglect of exposure duration, producing poor predictors of adverse consequences to ecological receptors. The understandable preoccupation of early ecotoxicologists, who wished to control point source pollutants, fostered this emphasis on intensity of exposure; exposure duration could be considered simply as "acute" or "chronic." Beyond point source control, there is little justification for this preoccupation. Despite arguments to do otherwise,[1] central metrics (e.g., LC50 and EC50 values) derived for acute or chronic scenarios have become "generally accepted as [the] conservative estimate of the potential effects of test materials in the field."[2]

Paracelsus never suggested that dose was more important to consider than exposure duration. Current ecological risk assessment requires more accurate predictions than afforded by temporally deficient effect metrics. Time to event methods provide one way of including both exposure intensity (dose or concentration) and duration in predictions of effect. As illustrated in this volume, more information is accumulated from standard tests if time to event methods are used instead of conventional methods. This additional information allows consideration of exposure duration and enhanced statistical power and model fitting. These statistical qualities of time to event methods allow ready inclusion of dose or concentration, producing a method for estimating effect based on intensity and duration of toxicant exposure.

There are nonparametric, semiparametric, and parametric time to event methods. All assume that a discrete event such as death, birth, stupefaction, diagnosis of cancer, or disease onset, occurs in time. Often, some individuals will remain unresponsive at the end of the experiment or period of observation. For example, times to death during exposure might be noted, with some individuals surviving beyond the end of the experiment. Surviving individuals are designated as censored because the observer only knows that the individual would have a time to death greater than the duration of the experiment. More specifically, they are right-censored because their times to death would be on the right side of a distribution of individual times to death. The only information available for such individuals is that they survived to the end of the experiment. So, time to event data sets contain times to death for many individuals and minimum times to death for others. Continuous or discrete covariates such as age, sex, exposure concentration, temperature, genotype, or season can also be incorporated into the record for each individual in the time to death data set. As an example, times to death could be noted for individual fish in a toxicity test with additional notation of fish wet weight upon death, exposure concentration, and sex. The time to death could be recorded precisely or as having occurred within the interval elapsed since the last observation, e.g., fish X died between 8 and 12 h of exposure.

Nonparametric time to event methods include conventional life or actuarial tables and Product-Limit methods.[3,4] With the Product-Limit methods, the cumulative survival at time, t is estimated based on the deaths occurring at all intervals up to and including t.

$$\hat{S}(t_i) = \prod_{j=1}^{i}\left(1 - \frac{d_j}{n_j}\right) \qquad 1.1$$

where i = the interval label i for t_i; n_j = the number of individuals alive and available to die just before time, t_j; d_j = the number of individuals dying at t_j. The variance for this estimate of cumulative survival is calculated with the following equation for all times prior to the end of the exposure.

$$\hat{\sigma}^2(t_i) = \frac{\hat{S}(t_i)[1 - \hat{S}(t_i)]}{N} \qquad 1.2$$

where N = the total number of exposed individuals. Obviously, the cumulative mortality can be estimated from the cumulative survival, $F(t_i) = 1 - S(t_i)$ and the estimated variance calculated above is applicable to $F(t_i)$ also.

Nonparametric tests for significant differences among survival curves generated for different classes, e.g., exposure concentrations or sexes, include log-rank and Wilcoxon rank tests.[5,6] Analogous to post-ANOVA testing for concentration effects, such results could be used to estimate a time to event NOEC, as shown in Chapter 2, if data were inappropriate for fitting to a conventional time to event model.

The Cox proportional hazard model is a semiparametric method for fitting time to event data.[4,7] In this model, the proneness to die or instantaneous mortality rate (hazard, $h()$) is modeled.

$$h(t, x_i) = e^{f(x_i)}h_0(t) \qquad 1.3$$

where $h(t, x_i)$ = the hazard at time, t, for an individual characterized with the value x_i for the covariate, x; $h_0(t)$ = the baseline hazard; and $e^{f(x_i)}$ = a function relating $h(t, x_i)$ to the baseline hazard. The $e^{f(x_i)}$ makes the $h(t, x_i)$ proportional to the baseline hazard at all exposure times. Regardless of exposure duration, the hazard of one class (e.g., females) is proportional to that of another (e.g., males). No attempt is made to fit a parametric model to the baseline hazard; instead, a function is fit to the baseline hazard and then attention is paid to fitting for proportionality of hazards, i.e., $e^{f(x_i)}$.

A fully parametric, proportional hazard model is produced if a specific model is fit to the $h_0(t)$. Common functions used to produce a proportional hazard model are the exponential and Weibull.

$$\text{Exponential : Survival Function} = e^{-at} \qquad\qquad 1.4$$

$$\text{Weibull : Survival Function} = e^{-(\alpha t)^{\beta}} \qquad\qquad 1.5$$

The $e^{f(x_i)}$ can take on many forms. A χ^2 test can be used to test for significant effect of a covariate in the model.

Often, hazards do not remain constant and a proportional hazard model cannot be used. In such a case, an accelerated failure time model is applied: Ln $t_i = f(x_i) + \varepsilon_i$ where ε_i is the error term associated with i and $f(x_i)$ can take on a variety of forms. Often, the error term is modeled with a log-normal, log-logistic, or gamma function.

Actual data fitting to accelerated failure time or proportional hazard models is most often done by computer. Right censoring requires an iterative maximum likelihood approach. However, because of the wide use of these methods, most commercial statistical packages do these analyses, e.g., SAS.[5] Fitting of time to death to a model that includes exposure concentration results in a model including exposure duration and intensity. Effects can then be estimated as a consequence of both, as described by Newman and co-authors in Chapter 3, and in other publications.[6,8–13]

1.3 Regulatory and scientific rationale for time to event methods

The brief overview given in the previous section has shown how interpretation of data might be enhanced by the use of TTE methods. But, is such an enhancement necessary? Current ecotoxicity databases are filled with median lethal or effective concentrations (LC50s or EC50s) and no observed effect concentrations (NOECs). Many regulatory guidelines and frameworks also require — or at least expect — these types of data. Although most environmental regulators would acknowledge that an LC50 is a rather crude measure of toxicity, it has the advantage of wide acceptance. Similarly, while the use of an NOEC to summarize toxicity data is widely criticized,[10] it is ubiquitous in method guidelines designed to test for sublethal effects of chemicals, and is favored over the LC50 by regulators as a starting point for extrapolating to "safe" chemical concentrations in the environment. It might also be argued that the application of large safety factors, from 10 to >1000, to any type of summary statistic derived from laboratory toxicity tests, means that a small increase in accuracy or precision from use of TTE approaches is largely meaningless (see Chapter 10 for further discussion of this).

We recognize that acceptance of TTE approaches as a *replacement* for current toxicity summaries is unlikely in the near future. Too much time and effort have been invested in generating LC/EC50s and NOECs for practitioners to abandon them overnight. Besides, despite manifest shortcomings, use of these summary statistics is better than no toxicity data at all when performing risk assessments. Evolution, rather than revolution, is the most likely method by which TTE methods will be adopted. Practitioners are likely to find that use of TTE does not preclude the calculation of traditional summaries (see Chapter 5), and that TTE offers them greater possibilities for interpreting their data.

Risk assessment for regulatory agencies is performed to enable the labeling and classification of chemical products, and the risk management of chemical products and contaminated media.[14] Decisions made by regulators about the risks posed from exposure to hazardous chemicals are often based on rather limited data sets, summarized as LC50 or NOEC values. These data are difficult to use if information is required on:

- likely biological responses to high concentrations of a chemical over a period shorter than that for reported toxicity values (e.g., during a pulsed pollution event in a river)
- likely biological responses to low concentrations of a chemical over a period longer than that for reported toxicity values (e.g., when a persistent compound is present in soil)
- likely effects of contamination on population viability, if data on mortality and reproduction are reported as NOECs, which cannot be used to parameterize demographic models

Such questions over the temporal component of toxicity and the relationship between individual and population toxicity are commonplace in almost any risk assessment.

Environmental regulators currently use large "safety factors" to compensate for uncertainties such as those outlined above. However, use of safety factors conflates all of the uncertainties surrounding a potentially hazardous chemical, making the risk assessment process more opaque than it need be. Some uncertainties will likely always remain during risk assessment, because of the great complexity of natural systems. However, use of data on time to event can eliminate or reduce gross uncertainties over the time course of toxicity, and the relationship between toxicity to individuals and populations. Using available data to reduce uncertainties such as these will allow regulators to concentrate their efforts and resources on other areas of the risk assessment process.

The job of scientists is to discover and describe underlying mechanisms in nature.[6] Although there is no one way to do this,[15] it is clear that ignoring the majority of data collected in an experiment and the many statistical techniques that have been successfully used in other disciplines is not the hallmark of good science. If environmental regulation is to be based on sound scientific principles, it is simply unacceptable for those involved in it to continue using methods of proven inferiority. At the very least, TTE data should be gathered and analyzed correctly alongside the generation of LC50, EC50 and NOEC summaries.

References

1. Sprague, J., Measurement of pollutant toxicity to fish. I. Bioassay methods for acute toxicity. *Water Research* 3, 793, 1969.
2. Parrish, P.R., Acute Toxicity Tests, in *Fundamentals of Aquatic Toxicology*, Rand, G.M. and Petrocelli, S.R., Eds., Hemisphere Publishing, Washington, D.C., 1985, 31.
3. Miller, R.G., Jr., *Survival Analysis*, Wiley-Interscience, New York, NY, 1981.
4. Cox, D.R. and Oakes, D., *Analysis of Survival Data*, Chapman & Hall, London, UK, 1984.
5. Allison, P.D., *Survival Analysis Using the SAS® System. A Practical Guide*, SAS Institute Inc., Cary, NC, 1995.
6. Newman, M.C., *Quantitative Methods in Aquatic Ecotoxicology*, CRC/Lewis, Boca Raton, FL, 1995.
7. Cox, D.R., Regression models and life tables (with discussion). *J. Royal Stat. Soc.*, B34, 187, 1972.
8. Dixon, P.M. and Newman, M.C., Analyzing Toxicity Data Using Statistical Models for Time to Death: An Introduction, in *Metal Ecotoxicology: Concepts and Applications*, Newman, M.C. and McIntosh, A.W., Eds., Lewis, Chelsea, MI, 1991, 207.
9. Newman, M.C., *Fundamentals of Ecotoxicology*, CRC/Ann Arbor, Boca Raton, FL, 1998.
10. Crane, M. and Newman, M.C., What level of effect is a no observed effect? *Environmental Toxicology and Chemistry*, 19, 516, 2000.
11. Newman, M.C. and Aplin, M.S., Enhancing toxicity data interpretation and prediction of ecological risk with survival time modeling: an illustration using sodium chloride toxicity to mosquitofish (*Gambusia holbrooki*). *Aquatic Toxicology*, 23, 85, 1992.
12. Newman, M.C. and Dixon, P.M., Ecologically Meaningful Estimates of Lethal Effect to Individuals, in *Ecotoxicology. A Hierarchical Treatment*, Newman, M.C. and Jagoe, C.H., Eds., CRC/Lewis, Boca Raton, FL, 1996, 225.
13. Newman, M.C. and McCloskey, J.T., Time to event analyses of ecotoxicity data, *Ecotoxicology*, 5, 187, 1996.
14. Whitehouse, P. and Cartwright, N. Standards for Environmental Protection, in *Pollution Risk Assessment and Management: A Structured Approach*, Douben, P.E.T., Ed., John Wiley and Sons, Chichester, 1998, 235.
15. Crane, M. and Newman, M.C. Scientific method in environmental toxicology. *Environmental Reviews*, 4, 112, 1996.

chapter 2

Time to event analysis of standard ecotoxicity data

Mark Crane and Albania Grosso

Contents

2.1 Introduction

The responses of organisms to toxicants depend upon both the intensity and duration of exposure. However, current approaches to the analysis of ecotoxicity data focus on identifying the concentrations of toxicant causing specific biological effects. The time course of toxicity is rarely considered in much detail.[1] This chapter describes some simple approaches that explicitly address the influence of time on toxic effect. It is written for ecotoxicologists and risk assessors with no mathematical expertise. The aim is to show the advantages of time to event analyses by taking some standard toxicity test results and analyzing them using traditional approaches to produce statistical summaries. These are the lethal or effective concentration (LC/ECx), where x is some suitable percentile, e.g., x = 50, and the no observed effect concentration (NOEC). The data are then analyzed using simple nonparametric time to event (TTE) approaches available in standard statistical software packages. A comparison of the two sets of results shows some of the advantages of even very simple TTE analysis for the risk assessment of chemicals.

2.2   Standard statistical analysis of ecotoxicity tests

Three main statistical approaches are available for analyzing ecotoxicity data: hypothesis testing to determine a NOEC, regression analysis to estimate a time-specific effect concentration, and time to event analysis.[2] Only the first two approaches are currently used routinely by ecotoxicologists.

The NOEC is the highest concentration in a bioassay producing a response that does not differ from the control when compared in a statistical significance test. It is usually calculated from concentration-response data by using analysis of variance (ANOVA), followed by a multiple comparison test, such as Dunnett's or Tukey's test.[3] This can only be done if there is replication at each concentration. The perceived advantage of the NOEC is that it is easy to understand.[2] However, the many disadvantages to its use have been discussed extensively elsewhere.[4–14]

The estimation of an ECx value overcomes many of the problems associated with hypothesis testing,[5] and is the usual form of analysis for acute ecotoxicity experiments. Data from fixed times of observation (usually 24, 48, 72 or 96 h in acute bioassays) are transformed so that least-squares fits can be made to linear models. Linearity is usually achieved by logging the exposure concentration and converting the response to its probit[15,16] or logit.[17] ECx values are then estimated for the magnitude of effect that interests the investigator. This is usually an EC50 or LC50, because more precise estimates are possible at this median point.

Although there is general agreement that the use of a dose-response curve to estimate an ECx has many advantages over the derivation of an NOEC, the calculation of an ECx at specific time intervals still uses data suboptimally. This is because most investigators will take some measurements during the course of a bioassay, especially if survival is the endpoint. These data from intermediate observation periods are often not used in the final estimation or risk assessment.

2.3   Time to event analysis

The importance of the duration of exposure in aquatic toxicity tests has been recognized for many years.[18,19] Traditionally, the standard analysis of exposure duration was to plot the mean or median survival time (or its reciprocal) against the toxicant concentration (or log concentration).[20–27] A line was then fitted, either by eye or by using a more formal model.[28] A "safe" level of chemical or effluent over an indefinite period was then estimated from this.[23] Sprague[27] recommended the lethal threshold concentration ("that level of toxicant beyond which 50% of the population cannot live for an indefinite time") as the most useful criterion of toxicity, and suggested a graphical method for its estimation. This involves plotting survival at each exposure concentration over successive time intervals until each response reaches an asymptote. The lethal threshold concentration is then estimated from these asymptotic responses. An alternative is to plot the toxicity curve of the median lethal concentration against time.[29]

More recently, Sun et al.[30] recommended the use of survival time modeling and accelerated life testing in ecotoxicity, an approach discussed in more detail in Chapter 4. Other researchers have also proposed survival time models as a method for integrating time, concentration, response, and ecologically important covariables such as organism weight and sex.[31-36] These approaches are commonly employed in engineering, where "failure time analysis" is used to analyze the failure of a component (see Chapter 9); medicine, where "survival analysis" is used to analyze the deaths of patients; and ecology, where "time to event" analysis can be used to analyze the arrival of a migrant or parasite, the germination of a seed, organism death,[37] or the time to a particular behavioral display (see Chapter 8). Parametric models can be based upon any theoretical distribution. For example, Newman et al.[35] made their choice from the normal, lognormal, Weibull, gamma and log-logistic distributions by comparing the fit of each of these to the data. Sun et al.[30] chose a Weibull distribution.

Nonparametric tools, such as life tables and the Kaplan-Meier method,[38] are also available for analyzing time to event data when the underlying distribution remains unknown. The life table approach groups occurrence times into intervals, while the Kaplan Meier, or "product-limit," approach produces an estimate at each individual occurrence time.[3,37,39,40] Calculations for product limit estimates of survival and the variance around these estimates are provided in equations 1.1 and 1.2 in Chapter 1. Survival curves can be compared using a range of different significance tests.[3,37,41-43] Both life tables and the product limit approach are used to estimate two important functions that describe the data: the survival function and the hazard function.

The survival function $S(t)$ is the probability of surviving longer than a particular time, t. The hazard function or rate $h(t)$ is the conditional probability density of an event, such as death, happening at time t, given that it has not occurred before then. It is also known as the "instantaneous failure rate," "proneness to fail" or "force of mortality"[3] and can be calculated by dividing the unconditional mortality probability density function $f(t)$ by the survival function $S(t)$:

$$h(t) = \frac{f(t)}{S(t)} \qquad\qquad 2.1$$

2.4   Time to event analysis of standard ecotoxicity data

Risk assessment frameworks are designed to help environmental regulators make rational decisions about how to protect the environment from harm without undue costs to humans. Risk assessment may be either predictive or retrospective,[1,49] and includes assessments of how new chemicals could impact the environment, or the extent to which pollutants already in the environment are causing adverse effects. Within both of these major classes of risk assessment, statistical models may be useful for prediction, description, explanation or extrapolation.[1,50] The first two examples in this section will deal with analyses of data from predictive toxicity tests with earthworms and fish. The final example analyzes data from a retrospective effluent test with waterfleas.

2.4.1   Predictive chemicals testing

Whitehouse and Cartwright[51] and Campbell and Hoy[52] describe the hazard quotient approach, which is currently the main way in which chemicals are assessed for their intrinsic hazard and likely risk to the environment. Predicted no effect concentrations (PNECs) are compared with predicted environmental concentrations (PECs) to derive a PEC/PNEC ratio. The PNECs are usually EC/LC50 data from short-term lethal "acute" studies, or NOECs from long-term sublethal "chronic" studies, with "safety factors" of usually between 10 and 1000 applied. The PECs are estimated from direct measurement, or models of varying complexity. If the PEC/PNEC is below a value of 1, then the risk is assumed to be low and no further testing is required. From this it can be seen that estimating a correct NOEC from the data in hand is of great importance, as it might influence whether a chemical is considered safe in the environment, or whether more testing or product prohibition will be required.

2.4.1.1   Earthworm survival

The first data set analyzed in this chapter represents the type of experiment most commonly performed by ecotoxicologists: a toxicity test with survival as an endpoint. This type of experiment can be performed with waterfleas over 48 hours, fish over 4 days, or any other organism that an experimenter wishes to use. In this case, earthworms were exposed to a substance at 0, 125, 250, 500, 1000 and 2000 mg/kg. Four replicates were set up at each concentration and the animals were tipped out of the soil and assessed for survival after 8, 14 and 21 days of exposure. The data are presented in Table 2.1.

Let us begin with a standard analysis of these data to produce LC values and an NOEC, the summary statistics that are most frequently used by risk assessors. Toxstat 3.4 can be used to analyze the data at each of the observation periods (Table 2.2). The different models give broadly similar results for each ECx at each time period. The fit of the three

Table 2.1 Data on Earthworm Mortality over a 21-Day Period

Treatment	Replicate	Survivors			
		Day 0	Day 8	Day 14	Day 21
Control	1	10	10	9	7
	2	10	10	9	9
	3	10	10	10	10
	4	10	10	9	8
125 mg/kg	1	10	10	10	8
	2	10	9	7	5
	3	10	10	10	9
	4	10	10	7	6
250 mg/kg	1	10	10	7	0
	2	10	9	9	8
	3	10	9	8	0
	4	10	10	8	1
500 mg/kg	1	10	9	1	0
	2	10	10	10	0
	3	10	10	0	0
	4	10	7	2	0
1000 mg/kg	1	10	10	1	0
	2	10	3	0	0
	3	10	5	0	0
	4	10	6	0	0
2000 mg/kg	1	10	0	0	0
	2	10	0	0	0
	3	10	0	0	0
	4	10	0	0	0

parametric models can be assessed by the χ^2 lack of fit statistic. The lower the statistic, the better the fit. With these data, the complementary log-log model is clearly the best among those that were tried. The log-log model also produces a fit that falls below the critical χ^2 value of 9.49 (found from statistical tables, or read from the Toxstat output). If none of the parametric models had been suitable, the non-parametric Spearman-Karber approach might have provided an adequate answer. With these data, an NOEC could be calculated at all time periods, using ANOVA followed by Dunnett's test. As one would expect, both the NOEC and the ECx values decline with length of exposure as more organisms succumb to the toxicant. The summary statistics from each period of observation can be plotted as a toxicity curve, a traditional approach to the incorporation of time. This would clearly show that the LC50 declines through time and does not reach an asymptote.

How would risk assessors currently interpret these results? The 21-d NOEC is 125 mg/kg of the substance. However, this "safe" level is the same

Table 2.2 ECx and NOEC Analysis of Earthworm Data (mg/kg)

Day	Type of Analysis	χ^2 Lack of Fit	EC10 and 95% CI	EC50 and 95% CI	EC90 and 95% CI
8	Probit	30.62	397	924	2151
			315-501	782-1092	1612-2871
	Logit	18.24	457	960	2021
			357-584	824-1120	1543-2646
	Log-log	7.97	467	1010	1653
			364-599	884-1154	1390-1965
	Spearman-Karber	—	—	949	—
				828-1088	
	NOEC = 500 mg/kg				
14	Probit	12.98	113	323	929
			86-147	271-386	699-1235
	Logit	11.61	123	335	916
			91-165	281-400	683-1228
	Log-log	4.04	113	373	801
			79-161	316-441	646-993
	Spearman-Karber	—	—	396	—
				319-490	
	NOEC = 250 mg/kg				
21	Probit	4.65	65	148	336
			50-84	126-173	262-431
	Logit	5.21	67	150	335
			51-88	128-175	259-433
	Log-log	1.13	59	163	314
			40-85	139-192	255-387
	Spearman-Karber	—	—	188	—
				156-226	
	NOEC = 125 mg/kg				

as the lower 95% confidence limit of the LC50 estimated by probit analysis, and is considerably higher than the LC10. In other words, the NOEC might cause 50% mortality and thus may not be a safe level at all. Interpretation of this data set is also rendered more difficult because of control mortalities during the last two observation periods. A risk assessor would probably divide this NOEC by at least 10 to obtain a PNEC of 12.5 mg/kg for comparison with a PEC. Since the toxicity curve does not reach an asymptote, a conservative risk assessor might decide that a safety factor of 100 or higher is more appropriate, producing a PNEC ≤ 1.25 mg/kg.

Let us turn now to a TTE analysis of these earthworm data. Figure 2.1 shows that the cumulative proportion of earthworms surviving the toxicant declines with both concentration and duration of exposure. Statistical comparison of these curves[40] gives a χ^2 of 26.21 (p < 0.001) when the curve for worms exposed to 250 mg/kg is compared with the control curve. The

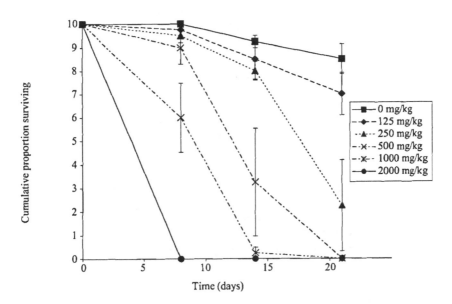

Figure 2.1 Cumulative number of earthworms surviving over 21 days. Error bars are standard errors of the mean.

curve for the worms exposed to 125 mg/kg does not differ at the 5% level, but there is an indication of a difference at around the 10% significance level (χ^2 = 2.568, p=0.107), suggesting that setting the NOEC at 125 mg/kg may be incautious.

When these data are analyzed to predict the mean life expectancy of the worms, the 500 mg/kg, 1000 mg/kg and 2000 mg/kg groups had life expectancies of 12.1 (± 95% confidence interval of 11.4 days), 7.7 (± 12.4) and 3.5 days (no CI could be calculated) respectively. Although the confidence intervals around these values are wide, this could be a useful way of communicating toxic risks to the public. Showing that organisms have a severely curtailed life expectancy when exposed to pollutants is likely to mean more to most people than statements about LC50 values and NOECs.

This first example shows that, even with rather minimal data from different time periods, it is possible to express the results in terms that might be more meaningful to the public.

2.4.1.2　Hatching of trout eggs

The next example involves more observations at different times, allowing use of the Kaplan-Meier approach, although in other respects the experimental design is rather poor. Trout eggs (50 per treatment) were placed in flow-through aquaria and exposed to 0, 1.79, 4.8 and 11.9 mg/L of a toxic substance. They were observed every day for 46 days for signs of hatching. The experimental units were not replicated. The data are plotted in Figure 2.2.

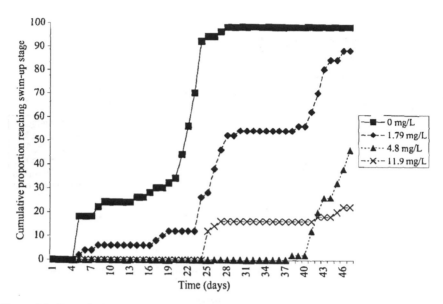

Figure 2.2 Cumulative percentage of trout eggs developing to swim-up stage.

Only three unreplicated concentrations and a control were tested in this experiment, so estimation of a standard NOEC is not possible with ANOVA. Under these circumstances, most practical ecotoxicologists would simply identify the NOEC as the concentration at the end of the experiment at which there was no apparent difference from the control: 1.79 mg/L in this experiment. The EC50 for these data also cannot be estimated with great accuracy or precision at the different time intervals. This is because of the low number of concentrations tested, the rather peculiar spacing of concentrations, and because more eggs hatched in the higher concentration than in the next lower concentration until near the end of the experiment.

How would risk assessors currently interpret these results? Because this is a "chronic" sublethal study they would probably be more concerned with the calculation of an NOEC than with an EC value. A risk assessor would divide the "apparent" NOEC of 1.79 mg/L by a safety factor of at least 10 to obtain a PNEC of 0.179 mg/L for comparison with a PEC.

Even this rather poorly designed experiment can be interpreted more fully than this. Statistical comparison of the survival curves gives a χ^2 of 38.65 (p < 0.001) when the curve for eggs exposed to 1.79 mg/L is compared with the control curve. Hence, the rate of egg hatching in 1.79 mg/L is significantly lower than in the controls. In nature, this might have important consequences. Eggs that hatch more slowly could be parasitized, predated or flushed away at a higher rate.

The hazard rate (Figure 2.3) clearly shows that hatching in controls has a maximum probability around the 20–30-day period. The maximum probability of hatching in the 1.79 mg/L exposure group occurs much later (around 40 days) and the probability of hatching in the two highest expo-

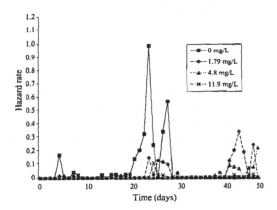

Figure 2.3 Estimated hazard rates for trout exposed to different toxicant concentrations.

sure groups remains rather low throughout the experiment. Some readers may be surprised that hazard rates can be above 1.0, which seems intuitively wrong. This is because hazard rates are instantaneous, rather than finite. The former can be converted to the latter using the equation: finite rate = 1.0 − $e^{\text{instantaneous rate}}$. In this example, on day 23, a control egg that has remained unhatched until this date has a $1 - e^{-1158} = 68.6\%$ probability of hatching within the next day. In contrast, an egg exposed to 1.79 mg/L has a $1 - e^{-0173}$ = 15.9% probability of hatching within the next day.

This second example shows another way in which TTE analysis can be used to express ecotoxicity results in terms that are meaningful to the public. Perhaps more importantly, it also shows how an NOEC established at the end of an experiment may not represent a truly safe concentration, especially if sensitive developmental stages have been delayed. TTE analysis can identify significant differences in temporal trends between treatments, allowing such developmental delays to be identified.

2.4.2   Direct toxicity of effluents

The final example in this chapter deals with data obtained from a retrospective study into the toxicity of an aqueous effluent. In the UK, the direct toxicity assessment (DTA) of aqueous effluents discharged to surface waters follows a similar approach to whole effluent testing (WET) developed in the United States.[53] Currently, controls are likely to be set on discharges using a PEC/PNEC approach very similar to that used to assess new chemicals. Samples of effluent are taken, diluted serially, and estimates of ECx and the NOEC made after the exposure of various test organisms for fixed periods of time. There has also been some debate about the use of limit tests, in which the response of organisms to a single concentration is compared with a control,[54] but this approach has not been adopted by the Environment Agency in England and Wales.

A TTE approach to analyzing DTA data has some potential advantages over traditional statistical approaches:

- The range within which NOEC and EC values can fall in an effluent bioassay is rather narrow (0–100% effluent). This distinguishes these types of tests from pure chemical tests in which, theoretically, the EC or NOEC could range from zero to a very high concentration. In other words, "accelerated testing" of environmental samples at higher concentrations than those actually found in the environment is not possible without artificial concentration of the samples. The use of more data points through the comparison of survival curves may help risk assessors to discriminate between effluent effects more effectively by providing more statistical power, and, hence, greater sensitivity.
- Effluent risk assessment can be difficult because the toxic components of a complex effluent are hard to distinguish. It may be possible to determine the broad classes of toxicants that cause biological effects by examining the rates of organism response through time.
- Limit tests based upon TTE use fewer organisms than concentration–response tests. This may be more ethical and cost-effective if data of similar value can be obtained.

2.4.2.1   *Waterflea reproduction*

A very simple example is presented here to illustrate the power of the TTE approach in DTA and WET programs. Waterflea reproduction has been used as a measure of effluent toxicity in the United States.[55] *Ceriodaphnia dubia* are kept separately in small chambers and exposed to effluent in replicates of 10 over a 7-day period. Time to first brood is one endpoint that can be measured. Data from two sets of waterfleas exposed to an effluent or a control, and observed every 8 hours over the duration of a test, are given in Table 2.3.

This minimal dataset cannot be used to produce an ECx or an NOEC as there is only one concentration and a control. There are no censored data; by the end of the experiment all of the animals have successfully produced a brood. How would an effluent risk assessor currently interpret these results? Because by the end of the test there is no difference in the success rates of the two treatment groups, it is very likely that the effluent would be passed as "safe." However, the survival curves in Figure 2.4 are significantly different (χ^2 = 9.206, p = 0.0024) because animals in the control group reproduce earlier than those exposed to effluent. In nature, this might put the offspring of the former group at a competitive advantage.

This final brief example shows how TTE analysis can potentially be used to reduce the extent and cost of effluent testing, and improve its sensitivity, while still providing data summaries of use to risk assessors.

Table 2.3 Time to First Reproduction for Waterfleas
Exposed to an Effluent or a Control

Treatment	Individual	Time to reproduction (hours)
Control	1	96
Control	2	104
Control	3	112
Control	4	112
Control	5	112
Control	6	120
Control	7	120
Control	8	120
Control	9	120
Control	10	128
Effluent	1	112
Effluent	2	120
Effluent	3	128
Effluent	4	128
Effluent	5	136
Effluent	6	136
Effluent	7	136
Effluent	8	144
Effluent	9	144
Effluent	10	144

2.5   Conclusions

TTE analysis has several advantages over more traditional approaches to data analysis. All of the data can be examined through time, either by comparing survival curves or by focusing on changes in survivorship and hazard at particular times during a test. The results of these analyses can be expressed as current probabilities or as projections of future probabilities, such as the expected life-span of an organism. TTE is not a cure for all of the data problems faced by environmental risk assessors. However, it does promote more careful interpretation of available data and helps to direct experimentation toward tests of biological significance and away from those that simply test statistical significance. The very simple approaches used in this chapter are accessible to anyone with a personal computer and do not require detailed statistical knowledge. Successful use may then lead on to

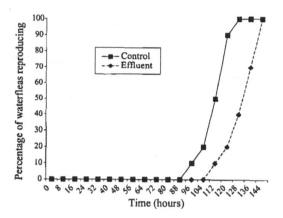

Figure 2.4 Percentage of waterfleas reproducing.

some of the more sophisticated, and potentially richer, approaches described in the next chapter.

Acknowledgements

We thank the Environment Agency and David Forrow, project leader, for supporting Project EMA 003, "The Statistical Analysis of Effluent Bioassays," upon which part of this chapter is based. Thanks also to Tim Sparks (CEH Monks Wood) for supplying the earthworm data. Comments from two anonymous referees substantially improved the quality of the chapter.

References

1. Suter, G.W. II, *Ecological Risk Assessment*, Lewis, Boca Raton, FL, 1993.
2. OECD, *Report of the OECD Workshop on Statistical Analysis of Aquatic Toxicity Data, Braunschweig, Germany, 15-17 October 1996*. Organization for Economic Cooperation and Development, Paris, France, 1997.
3. Newman, M.C., *Quantitative Methods in Aquatic Ecotoxicology*, Lewis, Boca Raton, FL, 1995.
4. Bruce, R.D. and Versteeg, D.J., A statistical procedure for modeling continuous toxicity data. *Environ. Toxicol. & Chem.*, 11, 1485, 1992.
5. Chapman, P.F., Crane, M., Wiles, J., Noppert, F. and McIndoe, E., Improving the quality of statistics in regulatory ecotoxicity tests, *Ecotoxicology*, 5, 169, 1996.
6. Crump, K.S., A new method for determining allowable daily intakes, *Fundament. & Appl. Toxicol.*, 4, 854, 1984.
7. Hoekstra, J.A. and van Ewijk, P.H., Alternatives for the no-observed effect level, *Environ. Toxicol. & Chem.*, 12, 187, 1993.
8. Kooijman, S.A.L.M., Parametric analyses of mortality rates in bioassays, *Water Research*, 15, 107, 1981.
9. Kooijman, S.A.L.M., An alternative for NOEC exists, but the standard model has to be abandoned first, *Oikos*, 75, 310, 1996.

9.   Kooijman, S.A.L.M., An alternative for NOEC exists, but the standard model has to be abandoned first, *Oikos*, 75, 310, 1996.
10. Laskowski, R., Some good reasons to ban the use of NOEC, LOEC and related concepts in ecotoxicology, *Oikos*, 73, 140, 1995.
11. Noppert, F., Leopold, A., and van der Hoeven, N., *How to Measure No Effect: Towards a New Measure of Chronic Toxicity in Ecotoxicology*. BKH Consulting Engineers, Delft, The Netherlands, 1994.
12. Pack, S., *A Review of Statistical Data Analysis and Experimental Design in OECD Aquatic Toxicology Test Guidelines*, Organization for Economic Cooperation and Development, Paris, France, 1993.
13. Skalski, J.R., Statistical inconsistencies in the use of no-observed-effect levels in toxicity testing, in *Aquatic Toxicology and Hazard Assessment: Fourth Conference*, Branson, D.R. and Dickson, K.L., Eds., American Society for Testing and Materials, Philadelphia, PA. 1981, 377.
14. Stephan, C.E. and Rogers, J.W., Advantages of using regression to calculate results of chronic toxicity tests, in *Aquatic Toxicology and Hazard Assessment: Eighth Symposium*, Bahner, R.C. and Hansen, D.J., Eds., American Society for Testing and Materials, Philadelphia, PA. 1985, 328.
15. Bliss, C.I., The calculation of the dosage–mortality curve, *Ann. Appl. Biol.*, 22, 134, 1935.
16. Finney, D.J., *Probit Analysis, 3rd Edition*, Cambridge University Press, Cambridge, UK, 1971.
17. Berkson, J., Application of the logistic function to bioassay. *J. Am. Stat. Soc.*, 39, 357, 1944.
18. Powers, E.B., The goldfish (*Carassius carassius*) as a test animal in the study of toxicity. *Illinois Biological Monographs*, 4, 127, 1917.
19. Bliss, C.I. and Cattell, M., Biological assay, *Ann. Rev. Physiol.* 5, 479, 1943.
20. Abram, F.S.H., An application of harmonics to fish toxicology. *Int. J. Air & Water Poll.*, 8, 325, 1964.
21. Abram, F.S.H., The definition and measurement of fish toxicity thresholds, in *Proc. 3rd Int. Conf., Advances in Water Pollution Research*, Munich, West Germany, September 1966, 1967, 75.
22. Alabaster, J.S. and Abram, F.S.H., Development and use of a direct method of evaluating toxicity to fish, in, *Proc. 2nd Int. Conf., Advances in Water Pollution Research*, Tokyo, Japan, 1964, 1965, 41.
23. Alderdice, D.F. and Brett, J.R., Some effects of kraft mill effluent on young Pacific salmon, *J. Fisheries Resources Board of Canada*, 14, 783, 1957.
24. Gaddum, J.H., Bioassays and mathematics, *Pharmacol. Rev.*, 5, 87, 1953.
25. Herbert, D.W.M. and Shurben, D.S., The toxicity of fluoride to rainbow trout, *Water and Waste Treatment* Sept/Oct 1964, 141.
26. Lloyd, R., The toxicity of zinc sulphate to rainbow trout, *Ann. Appl. Biol.*, 48, 84, 1960.
27. Sprague, J.B., Measurement of pollutant toxicity to fish. I. Bioassay methods for acute toxicity, *Water Research*, 3, 793, 1969.
28. Hey, E.N. and Hey, M.H., The statistical estimation of a rectangular hyperbola. *Biometrics*, 16, 606, 1960.
29. Heming, T.A., Sharma, A. and Kumar, Y., Time-toxicity relationships in fish exposed to the organochlorine pesticide methoxychlor, *Environ. Toxicol. & Chem.*, 8, 923, 1989.

30. Sun, K., Krause, G.F., Mayer, F.L., Jr., Ellersieck, M.R. and Basu, A.P., Predicting chronic lethality of chemicals to fishes from acute toxicity test data: theory of accelerated life testing, *Environ. Toxicol. & Chem.*, 14, 1745, 1995.

31. Diamond, S.A., Newman, M.C., Mulvey, M., Dixon, P.M. and Martinson, D., Allozyme genotype and time to death of mosquitofish, *Gambusia affinis* (Baird and Girard), during acute exposure to inorganic mercury, *Environ. Toxicol. & Chem.*, 8, 613, 1989.

32. Dixon, P.M. and Newman, M.C., Analyzing Toxicity Data Using Statistical Models of Time to Death: An Introduction, in *Metal Ecotoxicology: Concepts and Applications*, Newman, M.C. and McIntosh, A.W, Eds., Lewis, Chelsea, MI, 1991, 207.

33. Newman, M.C. and Aplin, M., Enhancing toxicity data interpretation and prediction of ecological risk with survival time modeling: an illustration using sodium chloride toxicity to mosquitofish (*Gambusia holbrooki*), *Aquat. Toxicol.*, 23, 85, 1992.

34. Newman, M.C., Diamond, S.A., Mulvey, M. and Dixon, P., Allozyme genotype and time to death of mosquitofish, *Gambusia affinis* (Baird and Girard) during acute toxicant exposure: a comparison of arsenate and inorganic mercury, *Aquat. Toxicol.*, 15, 141, 1989.

35. Newman, M.C., Keklak, M.M. and Doggett, M.S., Quantifying animal size effects on toxicity: a general approach, *Aquat. Toxicol.*, 28, 1, 1994.

36. Bedaux, J.J.M. and Kooijman, S.A.L.M., Statistical analysis of bioassays, based on hazard modeling, *Environ. & Ecol. Stat.*, 1, 303.

37. Muenchow, G., Ecological use of failure time analysis, *Ecology*, 67, 246, 1986.

38. Kaplan, E.L. and Meier, P., Nonparametric estimation from incomplete observations. *J.Am. Statist. Assoc*, 53, 457, 1958.

39. Cox, D.R. and Oakes, D., *Analysis of Survival Data*, Chapman and Hall, London, 1984.

40. Parmar, M.K.B. and Machin, D., *Survival Analysis, a Practical Approach*, John Wiley and Sons, Chichester, UK, 1995.

41. Pyke, D.A. and Thompson, J.N., Statistical analysis of survival and removal experiments, *Ecology*, 67, 240, 1986.

42. Pyke, D.A. and Thompson, J.N., Erratum, *Ecology*, 68, 232, 1987.

43. Hutchings, M.J., Booth, K.D. and Waite, S., Comparison of survivorship by the log-rank test: criticisms and alternatives, *Ecology*, 72, 2290, 1991.

44. Krebs, C.J., *Ecological Methodology*, Harper Collins, New York, NY, 1989.

45. Leslie, P.H., Tener, J.S., Vizoso, M. and Chitty, H., The longevity and fertility of the Orkney vole, *Microtus orcadensis*, as observed in the laboratory, *Proc. Zoolog. Soc. of London*, 125, 115, 1955.

46. West Inc., *Toxstat 3.4*, Western Ecosystems Technology, WY.

47. Unistat Ltd., *Unistat 4.5 User's Manual*, London, UK.

48 Stewart-Oaten, A., Rules and judgments in statistics: three examples, *Ecology*, 76, 2001, 1995.

49. Crane, M. and Newman, M.C., Scientific method in environmental toxicology, *Environ. Rev.*, 4, 112, 1996.

50. Barnthouse, L.W., The role of models in ecological risk assessment: a 1990s perspective, *Environ. Toxicol. & Chem.*, 11, 1751, 1992.

51. Whitehouse, P. and Cartwright, N. Standards for environmental protection, in *Pollution Risk Assessment and Management: A Structured Approach*, Douben, P.E.T., Ed., John Wiley and Sons, Chichester, 1998, 235.

52. Campbell, P.J. and Hoy, S.P., ED points and NOELs: how they are used by UK pesticide regulators, *Ecotoxicology*, 5, 139, 1996.
53. Tinsley, D., Johnson, I., Boumphrey, R., Forrow, D and Wharfe, J.R., The Use of Direct Toxicity Assessment to Control Discharges to the Aquatic Environment in the United Kingdom, in *Toxic Impacts of Wastes on the Aquatic Environment*, Tapp, J.F., Hunt, S.M. and Wharfe, J.R., Eds.. Royal Society of Chemistry, Cambridge, UK, 1996, 36.
54. Whitehouse, P., Crane, M., Redshaw, C.J. and Turner, C., Aquatic toxicity tests for the control of effluents in the UK — the influence of test precision, *Ecotoxicology*, 5, 155, 1996.
55. Sherry, J., Scott, B. and Dutka, B., Use of various acute, sublethal and early life-stage tests to evaluate the toxicity of refinery effluents, *Environ. Toxicol. & Chem.*, 16, 2249, 1997.

chapter 3

Applying time to event methods to assess pollutant effects on populations

Michael C. Newman and John T. McCloskey

Contents

3.1 Introduction

Ecological risk assessment aims to estimate the probability of some adverse consequence of contaminant exposure. Although ecological risk assessments involving endangered or exceptionally charismatic species do focus on individuals, effects most frequently of concern are those on populations. This being the case, it is incongruent that most information applied to estimating ecological risk comes from studies designed to assess effects on individuals. Fortunately, extrapolation from such information to population consequences can be improved by applying time to event methods in ecotoxicity testing and field surveys. Survival methods, including life table analysis, produce meaningful information (e.g., intrinsic rates of increase) amenable to projecting consequences such as local population extinction. The higher statistical power of the time to event approach relative to the conventional concentration-effect approach allows predictions incorporating important demographic and genetic covariates. Proportional hazard models allow easy estimation of relative fitness for genotypes that can be used in selection models to predict changes in tolerance or allele frequency through time. Our intent is to describe these advantages of time to event methods for predicting effects on population demographics and genetics.

3.2   *Importance of population consequences*

Methods for measuring pollutant effects focus on individuals despite the implied or stated goal of much regulation to protect species populations. "Protecting populations is an explicitly stated goal of several Congressional and Environmental Protection Agency mandates and regulations."[1] Current methods that are applied to assess effects on populations are derived without sufficient modification from mammalian toxicology, a discipline that focuses on protecting individuals. In our opinion, the prevailing use of these methods and data without additional modification to predict population effects results in compromised predictions.

> *There is an enormous disparity between the types of data available for assessment and the types of responses of ulti-mate interest. The toxicological data usually have been obtained from short-term toxicity tests performed using standard protocols and test species. In contrast, the effects of concern to ecologists performing assessments are those of long-term exposures on the persistence, abundance, and/or production of populations.[2]*

> *During the past two decades, toxicological endpoints (e.g., acute and chronic toxicity) for individual organisms have been the benchmarks for regulations and assessments of adverse ecological effects ... The question most often asked regarding these data and their use in ecological risk assessments is, "What is the significance of these ecotoxicity data to the integrity of the population?"[1]*

Why did the current practices become established in ecotoxicology? The uncritical adoption of methods from mammalian toxicology is not the only reason. These practices also have roots elsewhere. Predicting population consequences from physiological limits of individuals has a long history in ecology, the other major discipline contributing to ecotoxicology. For example, results from laboratory tests of individual tolerances to salinity and temperature extremes were applied to explain the distribution of species within estuaries,[3,4] and along latitudinal gradients.[5,6] Central paradigms such as Liebig's Law of the Minimum and Shelford's Law of Tolerance emerged from this general context.[7] Ecologists were accustomed to applying data for effects on individuals to predict those on populations. Perhaps this predilection resulted in less scrutiny than required during methods adoption.

Regardless, this reasonable ecological approach has been combined imperfectly with testing methods from mammalian toxicology to form the current ecotoxicological approach. Consequently, predictions of population effects must be made with suboptimal information. For example, a 96-h LC50 provides minimal understanding of the mortality to be expected through time for a population of *Daphnia magna*. Similarly, conventional reproductive data (the number of young produced by *Daphnia* that survive 21 days of exposure) are compromised because no information is available for reproduction by females that die during the test. Such unnecessarily incomplete mortality and natality data provide only gross insights into consequences for populations. There are tools amenable to predicting population effects based on appropriately measured effects on individuals. They require minor changes in current methods to produce better estimates of population effects and, with even further modification, could provide very sound information. Chief among these tools is time to event techniques.

3.3   Time to event methods

3.3.1   General description

Time to event methods draw from an experimental design in which groups of exposed individuals are monitored and the time until some event occurs (i.e., a discrete change in state[8]) is recorded for each individual. Often, the event is death, but other events relevant to population vital rates can be studied including time to reach swim-up stage, hatch, mature sexually, mate, spawn, flower, produce a brood, achieve an instar, or emerge as an adult. Exact times are recorded if practical, or the event may be noted as having occurred within an interval such as between 12 to 16 h.

One common aspect of time to event data is censoring. In the instance of a survival time experiment, individuals alive at the end of the experiment are censored; their times to death are known only to be longer than the duration of the experiment. Time to event data are often fit by maximum likelihood methods because of censoring.

Figure 3.1 is a diagram of the available time to event approaches. Like methods applied to calculating LC50 values, they encompass nonparametric, semiparametric, and fully parametric methods. Each can be used to fit time to event data, to predict times to events, or to test the statistical significance

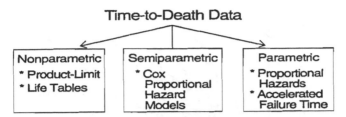

Figure 3.1. Time-to-event methods applicable to toxicity testing. Product-Limit (also called Kaplan-Meier) methods are nonparametric methods for estimating survival through time. Conventional life tables are another class of nonparametric techniques that have particular value in the analysis of populations. The semiparametric Cox proportional hazard model[28] assumes no underlying distribution for the mortality curve but does assume that the hazard (proneness to die during a time interval) remains proportional among groups such as males and females. Fully parametric methods have a specified distribution for the baseline mortality and specific functions describing the influence of covariates (e.g., exposure concentration) on time to death. Parametric models include accelerated failure time and proportional hazard models.

of covariates on time to event. A powerful tool is created by including exposure concentration (or dose) as a covariate in these time to event methods. With such a model, effect can be estimated as a function of both exposure intensity (concentration or dose) and duration.

3.3.2 General advantages of time to event methods

At first glance, these time to event methods may appear to offer little advantage to ecological risk assessors. This is not the case. There are eight reasons for this conclusion:

1. Exposure duration, a crucial determinant of exposure consequences, is explicitly included in these methods, thus affording accurate prediction of effect for different exposure durations. Explicit inclusion of duration is missing in the conventional approach (see Figure 3.2).
2. Including time to event information enhances the statistical power of tests because more data are extracted per test treatment, e.g., ten times to death per exposure tank vs. one proportion dead per exposure tank (see Figure 3.2). Numerous authors[9-12] point out the resulting increase in statistical power. All else being equal, time to event methods have higher statistical power than conventional methods because more data are collected from tests.
3. Because of their enhanced statistical power, time to event methods allow more covariates (e.g., toxicant concentration or water quality) to be included in predictive models. With low statistical power, the ability to test for significance for several covariates is very limited for conventional dose-response methods.

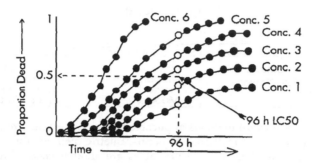

Figure 3.2. This figure illustrates the loss of information occurring if conventional methods are used instead of time to event methods. In a typical toxicity test, several exposure concentrations (six here) are established and the proportion dying of all individuals exposed to each is estimated at one time such as 96 h (open circles). If times to death were noted in each treatment, the temporal dynamics would have been captured and many more data points (all individuals dying) would have been available for analysis.

4. Because time to event data are generated for individuals, important qualities of individuals can be measured and included in predictive models. Examples include the influence of fish size[13,14] or sex[15,16] on sublethal or lethal effects.
5. Associated models allow expression of risks as probabilities.[12,14,16,17]
6. Associated methods allow many candidate models to be explored, resulting in accurate prediction. Candidate models are described in textbooks,[18–20] and ecotoxicological descriptions of time to event methods.[12,14,16,17,21]
7. Inclusion of time to event information does not preclude estimation of conventional endpoints such as the 96-h LC50.[12,16,21]
8. Results can be incorporated directly into well-established ecological, epidemiological, and toxicological models, e.g., demographic models combining survival information with effects on reproduction.

3.3.3   Advantages of time to event methods relative to population-level effects

Population consequences of exposure can be estimated with widely accepted demographic models if effects on mortality and natality were estimated through the lives of individuals. This is not possible with conventional ecotoxicity test data. For example, if the daily production of young (m_x) were known for *Daphnia* held at a specific toxicant concentration, no demographic analysis ($l_x m_x$ life table) could be completed because associated survival information would be available only as an LC50 (or some other LCx) at a specific time. (Often even this information is unavailable, as the *Daphnia* 21-day test does not provide reproduction data for adults dying during the test, natality data are available for survivors only.)

The dynamics of survival through time (l_x) would be needed to complete the life table. However, a complete $l_x m_x$ life table could be constructed if one could predict the proportion of individuals that survive to each day (l_x). Time to event methods generate such data. Important population qualities including the intrinsic rate of increase, age class-specific reproductive value and stable age structure could then be estimated. In the example used here, variation in mortality and natality among individual *Daphnia* could be used in stochastic simulations to generate the likelihood of local population extinction (see the RAMAS PC-based program for details[22]). Unlike the LC50 and NOEC information usually derived from ecotoxicity tests, rates of population increase and probabilities of local population extinction have clear ecological meaning.

3.4 Application to epidemiological description and prediction

Epidemiology is the study of disease incidence, prevalence and distribution in populations.[23] In the most rudimentary context, incidence can be expressed as a rate or cumulative incidence. Incidence rate is the number of individuals contracting or succumbing to disease relative to the study size, such as two smelter worker deaths from lung or nasal cancer in a study encompassing 1000 person years of exposure. The cumulative incidence is the number of people estimated to contract the disease in a given time, e.g., during the operational lifetime of the smelter or the average employment period for a worker. Prevalence is the incidence rate times the duration of exposure (e.g., two deaths/1000 person exposure years in the study) x 10, 000 exposure years = a prevalence of 20 worker deaths due to lung or nasal cancer.

To be most effective, epidemiological estimation requires that data for disease appearance be followed through time (duration of exposure).[23] In the above example involving nasal and lung cancer deaths of smelter workers, the duration of exposure for each worker should be known to describe fully the risk to that individual resulting from occupational exposure. The risk of dying from lung or nasal cancer increases as the exposure duration increases. Consequently, time to event methods are pervasive in environmental epidemiology and clinical science.[18,23] As an example, cancer-related deaths in the United States have been modeled through time based on age-, sex-, and race-dependent mortality rates.[24] Time-related deaths of asbestos textile workers were modeled with time to death techniques.[25] The compromised inclusion of time in ecotoxicity data inhibits the infusion of valuable epidemiological methods (e.g., time to event methods) to ecotoxicology. This makes it difficult to link laboratory test results to epidemiological surveys of pollutant-related effects in field populations. Time to event models could alleviate some of this difficulty, allowing calculation of effect probabilities or odds through time.

3.5 Population demographics: dynamics, structure and persistence

Demography is the quantitative study of birth, death, migration, age and sex in populations. Data for mortality through time are a major staple of demographic analysis. Current expressions of lethality, e.g., 96 h LC50 or NOEC, are inconsistent with demographic analysis and are ineffective for projecting from exposure to population consequences. But, as illustrated below, directly relevant information can be extracted from ecotoxicity tests with the methods outlined in Figure 3.1, and this information allows demographic analysis. To illustrate this point, survival alone is considered first, then reproductive qualities are included.

3.5.1 Survival

Survival during toxicant exposure can be modeled conveniently with time to death methods with important demographic covariates included in the models. For example, an accelerated failure time model was produced with the covariate fish size in addition to toxicant concentration.[14] These data were generated from a conventional toxicity test with duplicate tanks receiving seven concentrations of sodium chloride.[21] In the only deviation from convention, times to death for female mosquitofish (38 to 40 fish per tank) were noted at 8-h intervals for 96 h. Size (wet weight) was recorded as each dead fish was removed from a tank. The data included the salt concentration to which a fish was exposed, the duplicate tank from which it came, its wet weight, and its time to death. For survivors, times to death were recorded as greater than 96 h and treated as right censored in computations.

First, a nonparametric (product-limit) approach was used to estimate survival in duplicate tanks and to test for significant differences between duplicates (χ^2 from log-rank and Wilcoxon tests). Results from duplicate tanks were pooled during subsequent modeling because no significant differences were found. The accelerated failure time model selected for these data used natural logarithm transformations of toxicant concentration (g/L) and fish wet weight (g), and a log logistic distribution to describe the baseline survival curve.[16]

$$\ln TTD = \mu + \beta_s \ln[NaCl] + \beta_w \ln \text{weight} + \varepsilon \qquad 3.1$$

The fish's wet weight and the salt concentration to which it was exposed influenced its time to death. Or, rearranging Equation 3.1,

$$TTD = e^\mu e^{\beta_s \ln[NaCl] + \beta_w \ln \text{weight}} e^\varepsilon \qquad 3.2$$

where TTD = time to death, μ = intercept, β_s and β_w = coefficients for the effects of ln salt concentration and ln weight, respectively, and ε = an error term. Under the assumption of a log logistic model, $e^\varepsilon = e^{\sigma W}$ with σ = a scale parameter and W = a metameter for the logistic curve associated with some effect proportion, e.g., associated with p = 0.5 for the median time to death (TTD_{50}). Maximum likelihood methods were applied to estimate μ, β_s, β_w, and σ from these data. The fitted mosquitofish model was the following (Equation 3.3):

$$TTD_{50} = e^{15.211}\, e^{-4.1788\, \ln[NaCl]\, +\, 0.2659\, \ln\, weight}\, e^{0.2017\, W} \qquad\qquad 3.3$$

The W corresponding to any particular proportion of interest can be found in a table (e.g., Appendix Table 7 in Ref.12) or generated with a special function in many statistical or spreadsheet programs. Because W = 0 for the proportion of 0.5, the error term becomes $e^{\sigma 0} = 1$ for estimation of the median TTD. By changing W, predictions can be made for proportions other than 50% of the population dying. Predictions can be made for any combination of salt concentration and fish weight within the range used to produce the model. This model is particularly useful for predicting population consequences because individuals within natural populations vary in size. Also, the exact proportion of a specific size class that must die before the toxicant has an adverse impact on a population varies among species. The loss of large numbers of neonates due to a chemical release may (a small population of a long-lived, k-strategy species) or may not (a moderate sized population of r-strategy species) affect population persistence.

Conventional toxicity endpoints can still be estimated from survival models. For example, the 96-h LC50 can be estimated by rearranging Equation 3.1.

$$LC50 = e^{\dfrac{\ln 96\, -\, M\, -\, \beta_w\, \ln weight\, -\, 5w}{P_s}} \qquad\qquad 3.4$$

A 96-h LC50 of 11.26 is estimated using the average fish weight (0.136 g) in Equation 3.4. This is very close to that generated by applying the conventional trimmed Spearman-Karber method (11.58 g/L; 95% C.I. = 10.85 – 12.37) to the mortality data for 96 h. By changing the weight, time, or W in Equation 3.4, predictions can be made of LCx values for other proportions dying of various size classes after different exposure durations.

Survival could also have been summarized as a life table (l_x schedule) for these mosquitofish although information on size effects would have been lost (Table 3.1). Conventional demographic qualities such as life expectancies could then have been estimated from such tables and compared among treatments.

Individuals within populations differ in other important ways and the ability to estimate how these differences affect survival is essential in predicting consequences to populations. Sex-dependent mortality is one impor-

ble Analysis of Sodium Chloride-Induced Mortality of Mosquitofish

ʒ/L	(n = 76)	10.8 g/L	(n = 79)	11.6 g/L	(n = 77)	13.2 g/L	(n = 76)	15.8 g/L
	S. Error	S	S. Error	S	S. Error	S	S. Error	S
000	0	1.0000	0	1.0000	0	1.0000	0	1.0000
000	0	1.0000	0	1.0000	0	1.0000	0	1.0000
000	0	1.0000	0	1.0000	0	1.0000	0	0.8846
000	0	1.0000	0	1.0000	0	1.0000	0	0.5513
000	0	1.0000	0	1.0000	0	0.9605	0.0223	0.1923
000	0	1.0000	0	0.9870	0.0129	0.8289	0.0432	0.0513
000	0	0.9747	0	0.9610	0.0221	0.6974	0.0527	0.0256
342	0.0284	0.8987	0.0177	0.8571	0.0399	0.5658	0.0569	
079	0.0332	0.8734	0.0339	0.7403	0.0500	0.4342	0.0569	
816	0.0371	0.8101	0.0374	0.6883	0.0528	0.3026	0.0527	
421	0.0418	0.7722	0.0441	0.5974	0.0559	0.2368	0.0488	
289	0.0432	0.7722	0.0472	0.5714	0.0564	0.1579	0.0418	
026	0.0457	0.7722	0.0472	0.5325	0.0569	0.1053	0.0352	

e with the LIFETEST Procedure in SAS. S = estimated survival function (product-limit) similar to the l_x nal life tables; S. Error = standard error of the survival function estimated with Greenwood's formula.

Table 3.2 Results from Fitting a Cox Proportional Hazard Model to Mosquitofish Time to Death During Acute Exposure to Inorganic Mercury

Variable	Data Type	Estimate	Standard Error of Estimate	Calculated χ^2 Statistic	Probability[1] (df = 1 for all)
Tank	Categorical	0.1207	0.0861	1.965	0.1610
Sex	Categorical	–0.8033	0.0894	80.830	0.0001
$\log_{10}$ Weight	Continuous	–2.7551	0.2470	124.474	0.0001

[1]The probability of getting a χ^2 statistic as large as that calculated by chance alone.

tant example. Sex and size of mosquitofish exposed to mercury[26] and arsenic[27] have been effectively included in time to death models. In both cases, males died sooner than females and smaller individuals were more sensitive than larger individuals. Models such as Equation 3.1 were used to estimate these differences.

Notice that one of the variables (sex) in the above models was categorical, not continuous like wet weight. To include categorical or class information in these models requires an indicator or "dummy" variable. In the case of sex, one class (male) is selected as the reference sex and given an indicator variable value of 0: the remaining sex (female) is given an indicator variable value of 1. This allows a baseline survival curve to be constructed for males and adjusted by some factor, $e^{-\beta_s \cdot 1}$ where β_s = estimated β for the effect of sex on time to death and 1 = indicator variable for females. For males, an indicator variable of 0 results in no adjustment because e^0 is 1.

Newman and McCloskey[14] describe the modeling of such data for male and female mosquitofish exposed to dissolved mercury. Fish differing in size and sex were randomly placed into duplicate tanks; consequently, there were categorical (tank and sex) and continuous ($\log_{10}$ weight) qualities. A Cox proportional hazard model[28] was fit to these data (Table 3.2).

Note from Table 3.2 that there was no significant difference between exposure tanks but significant differences showed up between sexes and among fish of different sizes. The relative risks to males and females during inorganic mercury exposure were calculated.

$$\text{Relative Risk}_{\text{Male}} = e^{-0.8033(0)} = 1 \qquad\qquad 3.5$$

$$\text{Relative Risk}_{\text{Female}} = e^{-0.8033(1)} = 0.447 \qquad\qquad 3.6$$

Males were 1/0.447 or roughly twice as likely to die as were females. Using the Cox proportional hazard model in Table 3.2, the relative risk of dying could also have been estimated for male and female fish of different sizes. Such information would be extremely useful for projecting changes in natural populations during exposure to contaminants.

3.5.2 Survival and reproduction

A complete life ($l_x m_x$) table could be constructed for certain species if natality were noted through time in addition to survival. The example of collecting such data during a 21-day *Daphnia magna* test has already been given. This general approach was taken by several authors.[29-34] In most of these studies, survival (l_x = proportion surviving to the beginning of an age interval, x) and natality (m_x = number of offspring per female in age interval, x) were used in the Euler-Lotka equation to estimate r, the intrinsic rate of population increase.

$$1 = \sum_{x=0}^{x_{max}} l_x m_x e^{-rx} \qquad 3.7$$

This equation was solved with l_x and m_x values for each time (x) and some initial estimate of r. The process was repeated with adjustments to the estimated r upward or downward until the right side of the equation was approximately 1. The r associated with this final iteration was the authors' best estimate of the intrinsic rate of increase. This r can be interpreted as a population quality or, in the context of life history theory, a measure of the fitness of a group of individuals in a certain environment.[35]

Several important points emerged from these studies. Daniels and Allan[30] compared life table analysis of a copepod (*Eurytemora affinis*) and cladoceran (*Daphnia magna*) exposed to dieldrin and found that the population r appeared to be a better measure of toxicant effect than conventional endpoints. The r integrated various effects on survival and reproduction into a simple and meaningful endpoint. Another demographic study[36] made the important point that the most sensitive life stage of an individual may not be that most critical to population viability. This point was repeated in a slightly different context by Caswell,[37] "... it is not safe to assume that the most obvious effect of a toxicant on the vital rates is the source of that toxicant's effect on population growth rate." Caswell recommended demographic techniques to estimate toxicant effects to populations and described life table response experiments (LTRE) for this purpose. He also detailed the linkage of this life table approach to hazard models.[38]

As mentioned above, such data can be discussed relative to toxicant effects on individual life history traits[39,40] and associated theory used to interpret consequences to populations. To this end, a computer program (DEBtox) was developed to analyze data from standard aquatic toxicity tests based on changes in life history traits.[41] Modeling hazard through time was a central theme in this model.

3.6 Population genetics

These methods were used successfully to test for significant differences in times to death among genotypes exposed to various toxicants, including arsenate,[27] copper,[42] uranium[43] and inorganic mercury.[26,44,45] By applying a proportional hazard model to data from one of these studies,[26] Newman[12,14] converted relative risks for six glucosephosphate isomerase (GPI-2) genotypes to relative fitnesses of mosquitofish during acute mercury exposure (Table 3.3). To do this, relative risks of the six genotypes were calculated in a manner similar to that shown above for calculating relative risks of males and females (Equations 3.5 and 3.6). Next, the relative risk for the genotype with the lowest risk (0.402 for the 66/100 genotype) was divided by each genotype's relative risk to estimate relative fitness, e.g., 0.402/1.000 for the 38/38 genotype. These relative fitness values could then be used in conventional population genetics models to estimate the rate of allele change through time. Indeed, Newman and Jagoe[46] used these fitness values to determine the relative importance of mortality-driven accelerated drift and selection on the change in GPI-2 allele frequencies of populations exposed to mercury. They hypothesized that, during the proposed biomonitoring of population-level effects with this potential marker, accelerated drift caused by the mortality-driven reduction in effective population size could obscure any allele frequency changes due to selection. This hypothesis was not supported by simulations of GPI-2 allele frequency changes in mosquitofish populations.

More comprehensive genetic analysis of the changes in mosquitofish population genetics was made possible by including differences in reproduction among GPI-2 genotypes exposed to mercury. Mulvey et al.[47] measured differences in female sexual selection and fecundity selection for mosquitofish exposed to low concentrations of inorganic mercury. Relative to female sexual selection, the proportion of exposed females that were gravid differed among genotypes, with the 100/100 homozygote having the lowest proportion gravid of the six genotypes. The number of late stage embryos carried by each female also differed among genotypes; there was significant evidence of fecundity selection. These differences in fitness were combined

Table 3.3 The Conversion of Coefficients for the Categorical Information, Genotype, From a Proportional Hazard Model to Relative Fitness Values

GPI-2 Genotype	Estimated Coefficient	Relative Risk	Relative Fitness (w)
100/100	0.370	0.487	0.82
66/100	0.468	0.402	1.00
38/100	0.362	0.494	0.81
66/66	0.389	0.469	0.86
38/66	0.339	0.517	0.78
38/38	0	1	0.40

The reference genotype (38/38) has a coefficient set to 01 [modified from Reference 12].

in an individual-based model with those for differential survival to project changes in GPI-1 allele frequencies under several exposure scenarios.[46] With important qualifications, Newman and Jagoe[46] concluded again that the GPI-2 marker could be used as an indicator of a population-level response to inorganic mercury.

3.7 Conclusions

Although adequate to address questions posed when they were first established, current methods for generating and summarizing mortality data are inadequate for answering the complex questions associated with ecological risk assessment. Time to event methods have the potential to improve this situation. The two critical components of exposure intensity (concentration or dose) and duration, can be included in the associated predictions. This reduces uncertainty by eliminating the current necessity of predicting effect at different times by crude extrapolation from "acute" and "chronic" data. Equally important, results from time to event analyses can be used in models to predict effects on populations, i.e., models of population demographics and genetics. Prediction of effects on populations is crucial in ecological risk assessment because protection of species populations is the goal of key regulations. Uncertainty in projecting population fate during exposure to toxicants is reduced by appropriate application of time to event methods.

References

1. USEPA, *Summary Report on Issues in Ecological Risk Assessment*, EPA/625/3-91/018, United States Environmental Protection Agency, Washington, D.C., 1991.
2. Barnthouse, L.W., Suter, G.W. II, Roses, A.E. and Beauchamp, J.J., Estimating responses of fish populations to toxic contaminants, *Environ. Toxicol. & Chem.*, 6, 811, 1987.
3. Costlow, J.D. Jr., Bookhout, C.G. and Monroe, R., The effect of salinity and temperature on larval development of *Sesarma cinereum* (Bosc) reared in the laboratory, *Biol. Bull.*, 118, 183, 1960.
4. Kenny, R., Effects of temperature, salinity and substrate on distribution of *Clymenella torquata* (Leidy), Polychaeta, *Ecology*, 50, 624, 1969.
5. Hutchins, L.W., The bases for temperature zonation in geographical distribution, *Ecological Monographs*, 17, 325, 1947.
6. Dehnel, P.A., Rates of growth of gastropods as a function of latitude, *Physiological Zoology*, 28, 115, 1955.
7. Calow, P. and Sibly, R.M., A physiological basis of population processes: ecotoxicological implications, *Functional Ecology*, 4, 283, 1990.
8. Allison PD. 1995, *Survival Analysis Using the SAS7 System. A Practical Guide*, SAS Institute Inc., Cary, NC, 1995.
9. Gaddum, J.H., Bioassays and mathematics, *Pharma. Rev.*, 5, 87, 1953.
10. Finney, D.J., *Statistical Method in Biological Assay*, Hafner, New York, NY, 1964.

11. Sprague, J., Measurement of pollutant toxicity to fish. I. Bioassay methods for acute toxicity, *Water Res.*, 3, 793, 1969.
12. Newman, M.C., *Quantitative Methods in Aquatic Ecotoxicology*, Lewis, Boca Raton, FL, 1995.
13. Newman, M.C., Keklak, M.M., Doggett, M.S., Quantifying animal size effects on toxicity: a general approach, *Aquat. Toxicol.*, 28, 1, 1994.
14. Newman, M.C., McCloskey, J.T., Time to event analyses of ecotoxicity data, *Ecotoxicology*, 5, 187, 1996.
15. Mulvey, M., Keller, G.P., Meffe, G.K., Single- and multiple-locus genotypes and life-history responses of *Gambusia holbrooki* reared at two temperatures, *Evolution*, 48, 1810, 1994.
16. Newman M.C. and Dixon, P.M., Ecologically Meaningful Estimates of Lethal Effect in Individuals, in *Ecotoxicology. A Hierarchical Treatment*, Newman, M.C. and Jagoe, C.H., Eds., Lewis, Boca Raton, FL, 1996, 225.
17. Dixon, P.M. and Newman, M.C., Analyzing Toxicity Data Using Statistical Models of Time to Death: An Introduction, in *Metal Ecotoxicology: Concepts and Applications*, Newman, M.C. and McIntosh, A.W, Eds., Lewis, Chelsea, MI, 1991, 207.
18. Marubini, E. and Valsecchi, M.G., *Analyzing Survival Data from Clinical Trials and Observational Studies*, John Wiley & Sons, Chichester, UK, 1995.
19. Cox, D.R. and Oakes, D., *Analysis of Survival Data*, Chapman & Hall, London, 1984.
20. Miller, R.G., Jr., *Survival Analysis*, Wiley-Interscience, New York, NY, 1981.
21. Newman, M.C. and Aplin, M.S., Enhancing toxicity data interpretation and prediction of ecological risk with survival time modeling: an illustration using sodium chloride toxicity to mosquitofish (*Gambusia holbrooki*), *Aquat. Toxicol.*, 23, 85, 1992.
22. Ferson, S. and Akçakaya, H.R., *RAMAS/Age User Manual*, Exeter Software, Inc., Setauket, NY, 1991.
23. Ahlbom, A., *Biostatistics for Epidemiologists*, Lewis, Boca Raton, FL, 1993.
24. Monson, R.R., Analysis of relative survival and proportional mortality, *Computers & Biomed. Res.*, 7, 325, 1974.
25. Pearce, N., Checkoway, H. and Dement, J., Exponential models for analysis of time-related factors, illustrated with asbestos textile worker mortality data, *J. Occup. Med.*, 30, 517, 1988.
26. Diamond, S.A., Newman, M.C., Mulvey, M., Dixon, P.M. and Martinson, D., Allozyme genotype and time to death of mosquitofish, *Gambusia affinis* (Baird and Girard), during acute exposure to inorganic mercury, *Environ. Toxicol. & Chem.*, 8, 613, 1989.
27. Newman, M.C., Diamond, S.A., Mulvey, M. and Dixon, P.M., Allozyme genotype and time to death of mosquitofish, *Gambusia affinis* (Baird and Girard) during acute toxicant exposure: a comparison of arsenate and inorganic mercury, *Aquat. Toxicol.*, 15, 141, 1989.
28. Cox, D.R., Regression models and life tables (with discussion), *J. Royal Stat. Soc.*, B34, 187, 1972.
29. Marshall, J.S., The effects of continuous gamma radiation on the intrinsic rate of natural increase of *Daphnia pulex*, *Ecology*, 43, 598, 1962.
30. Daniels, R.E. and Allan, J.D., Life table evaluation of chronic exposure to a pesticide. *Can. J. Fisheries& Aquat. Sci.*, 38, 485, 1981.

31. Day, K. and Kaushik, N.K., An assessment of the chronic toxicity of the synthetic pyrethroid, fenvalerate, to *Daphnia galeata mendotae*, using life tables, *Environ.Poll.*, 44, 13, 1987.

32. Pesch, C.E., Munns, W.R. and Gutjahr-Gobell, R., Effects of a contaminated sediment on life history traits and population growth rate of *Neanthes arenaceodentata* (Polychaeta: Nereidae) in the laboratory, *Environ. Toxicol. & Chem.*, 10, 805, 1991.

33. Martínez-Jerónimo, F., Villaseñor, R., Espinosa, F. and Rios, G., Use of life tables and application factors for evaluating chronic toxicity of Kraft mill wastes on *Daphnia magna, Bull. Environ. Contam. & Toxicol.*, 50, 377, 1993.

34. Koivisto, S. and Ketola, M., Effects of copper on life-history traits of *Daphnia pulex* and *Bosmina longirostris, Aquat. Toxicol.*, 32, 255, 1995.

35. Stearns, S.C., *The Evolution of Life Histories*, Oxford University Press, Oxford, UK, 1992.

36. Kammenga, J.E., Busschers, M., Van Straalen, N.M., Jepson, P.C. and Bakker, J., Stress induced fitness reduction is not determined by the most sensitive life-cycle trait, *Funct. Ecol.*, 10, 106, 1996.

37. Caswell, H., Demography meets ecotoxicology: untangling the population level effects of toxic substances, in, *Ecotoxicology. A Hierarchical Treatment*, Newman, M.C. and Jagoe, C.H., Eds., Lewis, Boca Raton, FL, 1996, 292.

38. Caswell, H. and John, A.M., From the individual to the population in demographic models, in *Individual-based Models and Approaches in Ecology*, DeAngelis, D.L. and Gross, L.J., Eds., Chapman & Hall, New York, NY, 1992, 36.

39. Kooijman, S.A.L.M., Van Der Hoeven, N., and Van Der Werf, D.C., Population consequences of a physiological model for individuals, *Funct. Ecol.*, 3, 325, 1989.

40. Sibly, R.M., Effects of pollutants on individual life histories and population growth rates, in *Ecotoxicology. A Hierarchical Treatment*, Newman, M.C., Jagoe, C.H., Eds., Lewis, Boca Raton, FL, 1996, 197.

41. Kooijman, S.A.L.M. and Bedaux, J.J.M., *The Analysis of Aquatic Toxicity Data*, VU University Press, Amsterdam, The Netherlands, 1996.

42. Schlueter, M.A., Guttman, S.I., Oris, J.T. and Bailer, A.J., Survival of copper-exposed juvenile fathead minnows (*Pimephales promelas*) differs among allozyme genotypes, *Environ. Toxicol. & Chem.*, 14, 1727, 1995.

43. Keklak, M.M., Newman, M.C. and Mulvey, M., Enhanced uranium tolerance of an exposed population of the Eastern mosquitofish (*Gambusia holbrooki* Girard 1859), *Arch. Environ. Contam. & Toxicol.*, 27, 20, 1994.

44. Lee, C.J., Newman, M.C. and Mulvey, M., Time to death of mosquitofish (*Gambusia holbrooki*) during acute inorganic mercury exposure: population structure effects, *Arch. Environ. Contam. & Toxicol.*, 22, 284, 1992.

45. Heagler, M.G., Newman, M.C., Mulvey, M. and Dixon, P.M., Allozyme genotype in mosquitofish, *Gambusia holbrooki*, during mercury exposure: temporal stability, concentration effects and field verification, *Environ. Toxicol. & Chem.*, 12, 385, 1993.

46. Newman, M.C. and Jagoe, R.H., Allozymes reflect population-level effect of mercury: simulations of mosquitofish (*Gambusia holbrooki* Girard) GPI-2 response, *Ecotoxicology*, 7, 1, 1998.

47. Mulvey, M., Newman, M.C., Chazal, A., Keklak, M.M., Heagler, M.G., Hales, Jr., L.S., Genetic and demographic responses of mosquitofish (*Gambusia holbrooki* Girard 1859) populations stressed by mercury, *Environ. Toxicol. & Chem.*14, 1411, 1995.

chapter 4

Time–concentration–effect models in predicting chronic toxicity from acute toxicity data

Foster L. Mayer, Mark R. Ellersieck, Gary F. Krause, Kai Sun, Gunhee Lee, and Denny R. Buckler

Contents

1-56670-582-7/02/$0.00+$1.50
© 2002 by CRC Press LLC

4.1 Introduction

Both understanding and evaluating chronic toxicity of chemicals are essential to assessing their environmental hazards and making environmentally sound management decisions. Because of the large number and variety of industrial, agricultural and home-use chemicals released in the U.S. annually and the high cost and effort required for chronic tests, resources are often insufficient to obtain experimental information about long-term environmental impacts for all potentially hazardous chemicals. In comparison, acute tests are less costly and require less time for completion and, for these reasons, there is an abundance of acute toxicity data for numerous chemicals and organisms. Also, there are procedures for extrapolating effects data within classes of chemicals sharing similar chemical structures.[1] Thus, there is a strong rationale to relate acute and chronic toxicities of chemicals and to develop statistical and mathematical techniques to predict chronic toxicity based on data from acute toxicity experiments.

This chapter ties together classical methods (e.g., probit regression)[2] and time to event methods[3] to provide models that predict chronic toxicity from acute toxicity data. A significant interest exists in the use of short-term tests as a basis for linkage of exposure and time to response with chronic effects for ecological risk assessments. The ability to accurately and precisely associate chronic effects from acute time-concentration-effect data is a powerful approach that integrates various aspects of toxicokinetics and directly addresses a variety of uncertainties in terms of chronicity.

Using acute mortality data to estimate chronic toxicity (survival, growth, reproduction) to aquatic organisms customarily involves deriving an application factor[4] or an acute-to-chronic ratio,[5] both of which require acute and chronic toxicity testing. Kenaga[6] reviewed the principal measurements of the acute LC50, the maximum acceptable toxicant concentration (MATC), and the application factor (AF) used in determining chronic NOECs (the highest concentration causing 0% or no statistically significant effect) for many chemicals. The AF is derived by dividing the MATC for a compound, as determined in a chronic toxicity test with a given species, by the acute LC50 for the same compound tested with the same species. The acute-to-chronic ratio (ACR) is the inverse of the AF. The AF or ACR is then used to estimate chronic NOECs for other species for which only acute toxicity data exist.[7] These approaches have some limitations.

One limitation is that the biological end points and degrees of response are often not comparable between acute and chronic toxicity data. When one uses either the AF or the ACR, the acute median lethal concentration (LC50) is compared with the MATC, often derived from an endpoint other than mortality. Although different degrees of response (acute 50% vs. chronic no-effect) could be used when response slopes are similar, the slopes may be different. Additionally, the use of the AF or ACR method does not take into consideration the progression of mortality through time that is derived from acute toxicity tests. The concentration–time-response interaction has been addressed by Shirazi and Lowrie,[8] but they directed their efforts toward better defining the LC50. The acute toxicity value represents only one point in time (96-h LC50), and the relationship of degree of response with duration of exposure should be essential when one predicts chronic toxicity from acute toxicity data.

Lethality and other toxic effects are dependent on both concentration of a chemical to which an organism is exposed and length of exposure time. It is a common practice to investigate the toxicity of new and existing chemicals and effluents using acute toxicity tests. This is done by observing mortality resulting from exposure to a series of chemical concentrations, usually at 24, 48, 72, and 96 h. Time course distinguishes acute from chronic toxicity and also relates them as an integrated and progressive process. A time to response approach gives a better understanding of the progression of toxic effects over time, and survival time modeling has shown applicability in toxicological studies.[9,10]

Comprehensive alternative approaches are discussed in which simultaneous consideration is given to concentration, degree of response, and time course of effect, all of which are usually included in describing the results of an acute test, but are seldom used in hazard assessment. A consistent end point (mortality) and degree of response (0%) are used to predict chronic lethality from acute toxicity tests. One assumption is required: The concentration response relationship is a continuum in time.

4.1.1 Concept of acute and chronic toxicity

Statistical methods for evaluating acute toxicity, such as probit and logistic analyses, were well developed in the past few decades; however, alternative statistical concepts and methodology for chronic toxicity have not been fully explored. For acute toxicity, the LC50 (median lethal concentration) is in common use, and a number of methods for estimating the LC50 have been studied and widely used in toxicology.[11]

Both the experimental design used in acute studies and the random variable(s) observed should be understood. At least two alternative designs are considered.

Design I — This design uses $(k)(t)$ aquaria. Assume there are k concentrations to be used and there are t times of exposure when lethality will be observed. Each aquarium contains n test organisms. Observation times are randomized to the t aquaria within each concentration. The number dead are observed at time t associated with each of the k concentrations. All observations are independent.

Design II — Given k concentrations, the design utilizes k aquaria. Each aquarium uses n test organisms. Mortality is observed and "time to death" is recorded as the variable of interest. If the observation is at interval times (e.g., every 12 h), then exact time to death is not known. The methods presented accommodate this situation.

Important consequences of using these alternative designs require consideration. Design I is t times larger than Design II. It requires more data collection and time. Observations from Design I, mortality at an exposure time and concentration will be statistically independent. These data can be used without qualification for regression-type analysis. The resultant LC50 values (or other LC estimate) at each exposure time are independent. Percentage of mortality among exposure times, concentration being fixed, will be dependent when Design II is used. Mortality is monotonic and increases as exposure time increases. Any statistical problems associated with dependence are ignored when regression methods are used to estimate the joint effects of concentration and exposure time on lethality. Design II is used extensively, is more economical, and problems of dependence can be overcome.

The customary approach for summarizing chronic toxicity data in the United States is the use of MATC limits. The upper limit of the MATC is represented by the LOEL (the lowest-observed-effect level in a test). The lower limit is the NOEL (the highest no-observed-effect level in a test). The true chronic no-effect concentration is commonly believed to be within the limits of the MATC. The NOEL and LOEL are usually determined by statistical hypothesis testing. A number of researchers have pointed out the shortcomings of this concept and approach, and have suggested alternatives to the MATC.[12-15] The commonly defined MATC or no-observed-effect concentration is not an inherent characteristic or objective parameter of a chemical, but outcome of a statistical hypothesis test. A number of published data sets were used to illustrate that this method often overestimates the true no-effect concentrations of chemicals. In an acute toxicity test, one response endpoint (lethality) is usually analyzed, whereas in a chronic toxicity test, several response endpoints are often of interest (e.g., survival, growth and reproduction). Each such aspect of chemical chronicity (lethality, growth, or reduced fecundity) may significantly impact populations of organisms.

Every species has a natural spontaneous rate of mortality and consequently, in any chronic toxicity test, two competing risks exist: mortality caused by toxic effects of chemicals and spontaneous mortality. In an acute test, spontaneous mortality can be ignored since it is usually very small, but for a long-term exposure, mortality must be adjusted for competing risks. The appropriate probabilistic description of chronic lethality of a toxicant to a specific species is the change in natural mortality or the survival probability.

A no-effect (mortality) concentration, denoted by LC(r), at an acceptable risk level or negligible effect level $r\%$, is defined by the equation:

$$1 - r\% = S(T^*, \mathrm{LC}(r); \alpha, \beta)/S(T^*, 0; \alpha, \beta) \qquad 4.1$$

where $S(T^*, x; \alpha, \beta) = 1 - F(T^*, x; \alpha, \beta)$, $F(T^*, x; \alpha, \beta)$ is the mortality at exposure time T^* and concentration x, and $S(T^*, x; \alpha, \beta)$ is the survival probability at exposure time T^* and concentration x. Both α and β are parameter vectors. $r\%$ is a permissible percentage reduction in the test organisms' spontaneous survival proportion at exposure time T^*; $1 - r\%$ is the conditional probability of organisms surviving a toxicity test. Also, $r\%$ can be interpreted as extra risk, or mortality caused from other than natural death. Acceptable risk levels are usually specified as a small percentage (e.g., 0.01, 0.1 or 1%), which must be subjectively determined by individuals, organizations, or regulatory agencies. This definition is essentially the same as the concept of "safe doses" in carcinogenic studies.[16]

This concept is consistent with acute toxicity. The experiment duration T^* for an acute toxicity test is a short length of time (usually 24, 48, 72, or 96 hours). The survival probability of organisms in a clean environment during such a short period of time is approximately 1.0, that is, $S(T^*, 0; \alpha, \beta) = 1$ and $r\% = 1 - S(T^*, LC(r); \alpha, \beta) = F(T^*, LC(r); \alpha, \beta)$. Therefore, $LC(r)$ denotes the chemical concentration that is lethal to $r\%$ of the organisms at exposure time, T^*. The definition is the same as that used for acute toxicity testing.

Results from acute toxicity tests can thus be used to assess chronic lethality of chemicals. This involves two extrapolations — from high concentration to low concentration and from short-term to long-term exposure.

4.2 Methods

4.2.1 Two-step linear regression

Linear regression ($Y = a + bX$) was used to derive concentrations of lethal effect (LC0 = 0.01 %) for each observation time in an acute toxicity test and to predict the chronic no-observed-effect concentration (PNOEC) for lethality from those LC0s.[17]

4.2.1.1 Degree of response

In chronic toxicity tests, the NOEC is usually of greatest interest, whereas, in acute tests, the degree of response is usually 50%. Although a probit value[2] does not exist for 0 or 100%, an approximate value can be derived. In the use of probit analysis of acute toxicity data,[2,18] the probit value 99.99% mortality is used as an approximation of 100%. An approximate value for LC0 can thus be derived by subtracting the probit value for 99.99% (8.7190) from 10 to provide a probit value of 1.2810 for 0.01% mortality. Also, PNOECs derived using 0.01, 0.1, 1.0, or 5.0 % mortality as an estimate of zero become asymptotic to true zero at 0.01% mortality (Figure 4.1).

4.2.1.2 Time course of effect

Predicting chronic toxicity from acute toxicity data requires a means of estimating the LC0 for an indefinite period (chronic) from an acute toxicity test conducted over a finite period, usually 96 h or less. Green[19] and Sprague[20] provided approaches to the problem of estimating tolerance over an indefinite period, although they utilized the LD50 or LC50, respectively. They

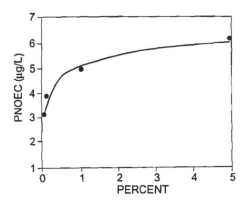

Figure 4.1 Relation of PNOECs (derived using 0.01, 0.1, 1.0, or 5.0% mortality as an estimate of zero) to true zero with kepone and fathead minnow data.[70] Reprinted with permission from F.L. Mayer, G.F. Krause, D.R. Buckler, M.R. Ellersieck, and G. Lee, 1994. Predicting chronic lethality of chemicals to fishes from acute toxicity test data: Concepts and linear regression analysis, *Environmental Toxicology and Chemistry*, Vol.13(4): 671–678. Copyright Society of Environmental Toxicology and Chemistry (SETAC), Pensacola, FL.

noted that, as the time of exposure becomes sufficiently long, the LD50 or LC50 approaches an asymptotic value. Green[19] suggested using a hyperbola to describe the relationship. A hyperbola can be expressed as a straight line by using the reciprocal of time (t) as the independent variable. The equation becomes LD50 = $a + b$ ($1/t$). Because $1/t$ approaches zero as t approaches infinity, the intercept (a) represents the LD50 over an indefinite time of exposure. The proposed method for estimating LC0 makes use of Green's approach for estimating time-independent LC50$_s$ from acute toxicity data.

4.2.1.3 Models

Acute flow-through toxicity tests should be conducted with strict adherence to standard test methods[21,22] to obtain estimates of LC0 over time. Periods of 24, 48, 72, and 96 h were selected because observations in standard acute toxicity tests are usually made at those time intervals. Data for observations less than 24 h were used when available, since these observations are very important in situations where most toxicity occurs during the early part of an acute test. The greatest concentration that causes no mortality and the least concentration that causes complete mortality were the concentrations used for 0 and 100% responses, respectively. All concentrations causing mortality ($0\% \leq x \leq 100\%$) were also included in our calculations. When regression analysis could not be conducted (less than three observations), the highest nonlethal concentration was used as the estimate of LC0 for that observation time. Having a range of mortalities for all time periods is best, although observation times with only 0 and 100% mortality values are acceptable if a concentration response is evident through time.

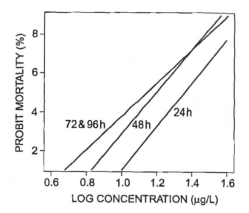

Figure 4.2 Dose-response curves used to derive the LC0 (0.01%) for various observation times in acute toxicity tests.[70] with kepone and fathead minnows $(1.281 = a + bX)$. Probit % mortality; 1.281 = 0.01%. 5.000 = 50%, and 8.719 = 99.99%. Reprinted with permission from F.L. Mayer, G.F. Krause, D.R. Buckler, M.R. Ellersieck, and G. Lee, 1994. Predicting chronic lethality of chemicals to fishes from acute toxicity test data: Concepts and linear regression analysis, *Environmental Toxicology and Chemistry,* Vol.13(4): 671–678. Copyright Society of Environmental Toxicology and Chemistry (SETAC), Pensacola, FL.

Linear regression analysis[23] was used to calculate the estimated LC0 at all observation times from acute flow-through tests (Figure 4.2) as percentage mortality (or probit percentage mortality) $= a + b$ (log concentration). The LC0s at each time period were then regressed against the reciprocal of time (Figure 4.3): LC0 $= a + b \, (1/t)$. The intercept (a) of this regression is the PNOEC for chronic lethality. Log transformations, log LC0 $= a + b \, (1/t)$ or log LC0 $= a + b \, (1/\log t)$, were required for better fits of data in 10 tests because of negative intercepts or curvilinear nature of the data. Confidence limits were not calculated and would be unrealistic because of the additive nature of combining the variation inherent in both steps. However, uncertainties (±2 standard errors of the estimate) were calculated and closely approximate 95% confidence limits.

4.2.2 Multifactor probit analysis

This method[24] is based on the concept of Mayer et al.[17] and uses several statistical models that simultaneously evaluate relationships among chemical concentration, time, and probit mortality to predict chronic response. The data required are quantal responses (i.e., mortality) observed for a particular species, with multiple observation times and at several concentrations of chemical. Data quality must be sufficient to permit estimates with small standard errors; hence, two minimum conditions are recommended. At least five concentrations or doses resulting in mortalities > 10% and < 90% over a fixed exposure time are desirable. The range in times

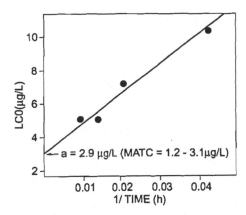

Figure 4.3 Prediction of the PNOEC value for lethality from acute toxicity test data[70] with kepone and fathead minnows ($LC0 = a + b[1/t]$). The intercept (*a*) represents the LC0 (2.9 µg/L) over an indefinite exposure time (PNOEC), and the maximum acceptable toxicant concentration (MATC) for chronic lethality was between 1.2 and 3.1µg/L. Reprinted with permission from F.L. Mayer, G.F. Krause, D.R. Buckler, M.R. Ellersieck, and G. Lee, 1994. Predicting chronic lethality of chemicals to fishes from acute toxicity test data: Concepts and linear regression analysis, *Environmental Toxicology and Chemistry*, Vol.13(4): 671–678. Copyright Society of Environmental Toxicology and Chemistry (SETAC), Pensacola, FL.

of exposure should be adequate to permit estimation of the trend in lethality within a concentration. The models will function with data that are below stated minimal requirements; however, accuracy of estimates may be diminished.

4.2.2.1 Models

The relationship between lethality of a chemical as a function of concentration and exposure time has been recognized for many years. Discussions and assessments are given by Bliss[25] and other scientists.[5,8,10,26,27]

Two basic models were formulated to represent the response surfaces of probit (percent) mortality as a function of toxicant concentration and acute exposure time: (1) probit $(P) = \alpha + \beta C_x + \delta T_x$ and (2) probit $(P) = \alpha + \beta C_x + \delta T_x + \gamma C t_x$, where P = probability of death. Parameters estimated are α, β, δ, and γ. These models allow C_x and T_x to assume several representations: C_x may be actual concentration or the logarithm of concentration; T_x is a representation of acute exposure time, its reciprocal, or logarithm. These two basic models can take several forms depending on how concentration and time are represented. The first model assumes that the derivative of probit (P) with respect to C_x is constant at each acute exposure time (parallelism). When nonparallelism exists or is assumed, the second model should be used. In that case, the partial derivative of probit (P) with respect to C_x is $\beta + \gamma T_x$, which allows for a change in slope as exposure time varies.

The two basic models are classified into three pairs, depending on the form of C_x and T_x; the first of each pair is a form of model 1 and the second represents model 2. The pairs and models are as follow:

Pair A. Two independent variables (log dose and log time).
  A1. Probit $P = \alpha + \beta \log (C) + \delta \log (T)$
  A2. Probit $P = \alpha + \beta \log (C) + \delta \log (T) + \gamma(\log (C) \log (T))$
Pair B. Two independent variables (log dose and reciprocal of time).
  B1. Probit $P = \alpha + \beta \log (C) + \delta/T$
  B2. Probit $P = \alpha + \beta \log (C) + \delta /T + \gamma(\log (C))/T$
Pair C. Two independent variables (log dose and reciprocal of log time).
  C1. Probit $P = \alpha + \beta \log (C) + \delta/\log (T)$
  C2. Probit $P = \alpha + \beta \log (C) + \delta /\log (T) + \gamma(\log (C)/\log (T))$

Standard error estimates for predicted chronic no-observed-effect concentrations (PNOEC) were computed using Fieller's rule,[28] because LCp is a ratio of random normal variables. All calculations were accomplished using the software program Multifactor Probit Analysis.[29]

A method was developed to compute maximum likelihood estimates of the parameters in each model studied. Response is quantal, and a normal tolerance distribution to the logarithm of chemical concentrations is assumed. When the normal tolerance distribution and probit transformation are utilized, a straight-line relationship results. This is important for small probit values (see[28], sec. 17.14). This relationship does not hold for some other tolerance distributions (i.e., logistic or angle). The resultant normal equations were solved numerically using an accelerated convergence scheme.

The estimates of α, β, δ, and γ, when required, provide an estimate of the equation of the surface, probit$(P) = (C_x, T_x; \hat{\alpha}, \beta, \delta, \hat{\gamma})$, and the caret indicates an estimate of an unknown parameter. When probit P is equal to a constant (the chosen tolerable constant fraction responding), then the concentration causing that level of response as a function of T_x can be obtained. For example, if probit $(P) = \alpha + \beta\log C + \delta/\log T$, then $\log (LCp) = ($probit $(P_c) - \alpha - \delta/\log (T))/\beta$. When probit $(P) = \alpha + \beta \log C + \delta T$, then $\log (LCp) = ($probit $(P) - \alpha - \delta T))/\beta$; $\log (LCp)$ is the logarithm of the lethal concentration yielding mortality p.

The concentration predicted to cause mortality probability P is found by substituting the desired chronic exposure time, T_{xc}, for T_x in the log (LCp) equation. Finally, the antilog gives the desired concentration.

When P is 0.0001, then probit $(0.01\%) = 1.281$, or a very close approximation of zero in the probit scale.[17] Other choices of P for chronic mortality could be used (i.e., 0.1%, 1%, etc.). The concentration LC 0.01 is used as the predicted concentration that has the effect of causing 0.01% mortality when exposure time is T_{xc}, and this value is used to calculate the PNOEC.

Associated with each PNOEC estimate is a heterogeneity factor (HF)[2]:

$$\text{Heterogeneity factor} = \sum_{i=1}^{k} \frac{(x_i - n_i P_i)^2}{n_i P_i (1 - P_i)(k - m)} \qquad 4.2$$

It is a chi-squared variable divided by degrees of freedom $(k - m)$; $k =$ the number of concentration-time combination data points, $m =$ number of parameters estimated; n_i, x_i, and $n_1 P_1$ are the number of organisms tested at point i, number responding, and the expected number responding, respectively. All HFs can be compared with 1.0 since the numerator is approximately a chi-squared variable with the mean equal to the denominator. Any HF > 1.0 is above average and $(k - m)$(HF) can be used to test for significant deviations from the model. If $(k - m)$(HF) exceeds the approximate tabular χ^2 value (5%), the methods of Mayer et al.[17] or Sun et al.[30] should be used. The smallest HF among the models tried using the same data is identified as min (HF) and is used as the criterion to choose among models.

The smaller the heterogeneity factor, the better the model fits the data. When comparing two models, the model with the smallest heterogeneity factor should be chosen. Generally, the size of the ±2 SE interval increases as the number of parameters in the model increase, while the heterogeneity factor may or may not increase. The minimum ±2 SE interval should not be used for model selection.

4.2.3   Accelerated Life Testing

Accelerated life testing theory was originally developed for and is commonly used in industrial reliability studies[31] (see Chapter 9). This results in shorter lives than would be observed under normal conditions. Such investigation places the test units (equipment, electronic components, etc.) under conditions that are more severe than normal. Accelerated life tests are used with short-term laboratory results as a predictor of the reliability under long-term standard stress.

This theory was applied to biological testing (i.e., short-term = acute, long-term = chronic), and extended to chronic toxicity.[30] Stress is defined as a function of concentration of a chemical, the test items as organisms, and failure as death of experimental organisms. When Design II is used and "time to death" of organisms by concentration is observed, methods for accelerated life tests apply under the assumption that the failure mechanism at all concentrations is the same (i.e., the same mode of toxicant action).

Statistical theory and methodology have been well developed for accelerated life testing in areas such as reliability and survival analysis. In reliability, the prime interest is in life distributions under normal stress. In many biological applications, similar problems involve estimating the relationship between life and variables that affect life. One way to model accelerated life testing data is through a model in which lifetime has a distribution that depends on stress variables. This model specifies the distribution of lifetime (*T*) given stress level *x*.

4.2.3.1 Models

The cumulative distribution function for failure time (*t*), $F(t, x; \alpha, \beta)$, involves a stress factor or concentration level, *x*. If *x* = 0, then a normal condition is assumed. The survival function is defined as $S(t, x; \alpha, \beta) = 1 - F(t, x; \alpha, \beta)$. Let $f(t, x; \alpha, \beta)$ be the probability density function for time to death. The hazard rate function specifies the instantaneous rate of failure at *T* = *t*, conditional on survival to time *t*, and is defined as $h(t, x; \alpha, \beta) = f(t, x; \alpha, \beta)/S(t, x; \alpha, \beta)$.

Proportional hazard models in which factors related to mortality have a multiplicative effect on the hazard function, play an important role in accelerated life testing theory. Under the proportional hazards assumption, the hazard function of *T*, given *x*, is of the product form:

$$h(t, x; \alpha, \beta) = g(x; \alpha)h_0(t; \beta) \qquad 4.3$$

where $h_0(t; \beta)$ is the spontaneous hazard rate, and the effect of a change in the concentration level *x* on spontaneous hazard rate h is a multiplication by a factor $g(x; \alpha)$ applicable to the full time range; $g(x;\alpha)$ is called the covariate function or the concentration effect function. The form of $h_0(t; \beta)$ is usually determined by knowledge about spontaneous life distributions. The key step to modeling such accelerated life tests is determining how to specify the form of $g(x; \alpha)$. Researchers in different scientific areas specify the effect function in various ways. For example, $g(x; \alpha) = \exp(\alpha x$,[32] is widely used in clinical trial data analysis and $g(x; \alpha) = \Sigma \alpha_s x^s$,[16,33] which is called the multistage model, is commonly used in carcinogenic experiments. Some specific covariate functions have been proposed in toxicology studies.[3,9,10,34] A form of $g(x;\alpha)$ suitable to summarize and interpret toxicity data follows.

Cox,[32] Kalbfleisch and Prentice,[31] and Lawless[35] suggested that time-dependent regression variables can be used in extending proportional hazard models. Mayer et al.[17] and Shirazi and Lowrie[8,36] also pointed out that concentration-response curves alone are not sufficient for studying toxicity of chemicals, and more comprehensive approaches, taking into account the progression of lethality through time, are needed to make full use of acute toxicity data. Following this reasoning, we used a time-dependent concentration effect function as follows:

$$g(x, t; \alpha) = 1 + ax^b t^c \qquad 4.4$$

where = (*a*, *b*, *c*) is a parameter vector; *a*, *b*, *c* > 0. The scale parameter (or scale factor) is *a*, and *b* and *c* are shape parameters (or shape factors) because the shapes of function curves depend on them. Note that, if *x* = 0, *g*(*x*) is 1, the corresponding survival probability becomes the spontaneous survival probability. The survival probability at concentration level x and time t will be:

$$S(t, x; \alpha, \beta) = \exp\left[-\int_0^t [1 + ax^b u^c] h_0(\mu; \beta) d\mu\right] \qquad 4.5$$

where μ represents an estimate of spontaneous (or baseline) death. Adjusting for competing risks (i.e., using the conditional probability of organisms surviving at concentration x and time t given that the organisms are surviving at time t in a clean environment), a concentration–time-response relation is obtained as follows:

$$Q(t, x; \alpha, \beta) = S(t, x; \alpha, \beta)/S(t, 0; \alpha, \beta) = \exp\left[-\int_0^t ax^b \mu^c h_0(\mu; \beta) d\mu\right] \qquad 4.6$$

This model assumes that both concentration levels and exposure durations have effects on survival probability or the fraction of responses, and hence, it has the ability to summarize the entire concentration–time-response data of a test. It also takes into account the spontaneous survival probability, and is suitable to describe both acute and chronic lethality data. Q is a survival function, including competing risks such as contaminant exposure.

The choice of life distribution is based on empirical knowledge. For the Weibull distribution, $h_0(t; \beta) = \lambda\gamma(\lambda t)^{\gamma-1}$, where λ and γ are parameters, then for $0 < t < T^*$, model (4.6) becomes $Q(t, x) = \exp\{-a^* x^b t^{c^*}\}$, where $a^* = \lambda a\gamma \gamma/(c + \gamma)$ and $c^* = c + \gamma$. For simplicity of notation, we denote a^* and c^* by a and c, respectively, and obtain the following model:

$$Q(t, x) = \exp\{-ax^b t^c\} \qquad 4.7$$

This model is a special case of model 4.6. For a fixed point in time, this model becomes a frequency distribution of tolerance, and, for a fixed concentration, the model is an ordinary survival probability distribution. In terms of log concentration, this tolerance distribution is the extreme value distribution widely used in the analysis of toxicity data without a time dimension.[37] In addition to the plausible derivations, another merit of this model is that its parameters provide some biologically meaningful explanation of toxicity. The scale factor a is a measure of the strength of toxic action. A large value of factor a indicates a chemical is very potent and rapid in action. More specifically, a is the value of –log (survival probability) at x = 1 and t = 1. The shape factors b and c, which describe the concentration-response and time-response surface shapes, respectively, are measures of the mode of response. Organisms that are very sensitive to incremental change in the concentration (or duration of exposure) have a large value of b (or c). In terms of multiple regression, log (a) is an intercept, and b and c are slopes of log concentration and log time, respectively.

The least-squares estimates of parameters of this model can be calculated directly from the data by using the multiple regression equation:

$$\log\left(-\log(Q(t, x))\right) = b(\log)x)) + c(\log(t)) + k \qquad 4.8$$

where $k = \log(a)$. However, the error term in this multiple regression is not normally distributed, and no confidence intervals for parameters can be obtained. The least-squares estimates are easy to calculate, but they are often inefficient.[38] To obtain more accurate estimators, a sophisticated procedure is employed. Statistical theory shows that maximum likelihood estimators are consistent, efficient and normally distributed in large samples, and thus, the maximum likelihood method was relied upon. The likelihood function for continuous data is:

$$L = \prod_{j=1}^{n}\prod_{i=1}^{m}[f(t_{ij}, x_j)]^{\delta_{ij}}[S(t_{ij}, x_j)]^{1-\delta_{ij}} \qquad 4.9$$

where $\delta_{ij} = 0$ if t_{ij} is censored and $\delta_{ij} = 1$ if t_{ij} is the time of death; $x_j, j = 1, ...,$ n are n concentrations. Although observing survival times in toxicity tests has merits, it is laborious and done infrequently. The commonly used method in acute tests is to record test mortality at selected times, usually 24, 48, 72 and 96 h. In a toxicity test, suppose that test organisms are inspected only once. If a response (e.g., death) is found, one knows only that death occurred before its inspection time. Similarly, if a response is not found, one knows only that its death time is beyond its inspection time. Such inspection data are called quantal-response data. However, as mentioned above, organisms are usually inspected more than once in an acute toxicity test. If an organism is found dead, one knows that the death occurred between that inspection time and the previous one. Also, if an organism is alive on its last inspection, one knows only that its death is beyond the inspection time. Such periodic data are called interval or grouped data, which provide more information than quantal-response data.[39] The likelihood function for interval data is:

$$L = \prod_{j=1}^{n}\prod_{i=1}^{m+1}[S(x_j, t_{i-1}) - S(x_j, t_i)]^{r_{ij}} \qquad 4.10$$

where $x_j, j = 1, ..., n$ are n exposure concentrations, $t_i, i = 1, ..., m$ are m inspection times; $0=t_0 < t_1... < t_m < t_{m+1} = \infty$, r_{ij} are the numbers of deaths observed in the interval (t_{i-1}, t_i) at concentration $x_j, i = 1, ..., m, j = 1, ..., n$; and $r_{m+1, j}$ are the numbers of surviving organisms at test termination time t_m and at concentration x_j. For model 4.7, the log likelihood function is:

$$\log(L) = \sum_{j=1}^{n} \sum_{i=1}^{m+1} r_{ij} [\log[e^{-\alpha(x_j)^b (t_{i-1})^c} - e^{-\alpha(x_j)^b (t_i)^c}]] \qquad 4.11$$

The Newton-Raphson algorithm or Quasi-Newton method can be used to find the maximum likelihood estimates of the parameters. The asymptotic variances of the estimators can be computed by using the observed Fisher information matrices,[35] but the computation is somewhat cumbersome.

To develop confidence intervals of parameters, large-sample procedures based on maximum likelihood were used. One such procedure is based on normal approximations to the distributions of the maximum likelihood estimates, and another one is a likelihood ratio method. The first method is to use the approximation:

$$(\hat{a}, \hat{b}, \hat{c}) \sim N[(a, b, c), I^{-1}] \qquad 4.12$$

where I is the observed information matrix (3×3). For $0 < p < 1$, the estimate of LC($100p$), the concentration that is lethal to $100p$ percent of the test organisms, is calculated by the formula:

$$x_p = \exp[(\log(-\log(1-p)) - \log(a) - \hat{c}\log(t))] \qquad 4.13$$

To compute the variance of x_p, asymptotic variance formulae of Rao[40] were used.

The second method is a likelihood ratio procedure. This method constructs a test of hypothesis and then converts the test to a confidence set. Consider the hypothesis H_0: LC($100p$) $= x_0$ versus H_1: LC($100p$) $\neq x_0$. Under H_0, $x_0 = \exp[(\log(-\log(1-p)) - \log(a) - c\log(t))/b$ or $a = -\log(1-p)/(x_0^b t^c)$. To find the maximum likelihood estimates, $\tilde{a}$, $\tilde{b}$ and $\tilde{c}$ under H_0, one substitutes $a = -\log(1-p)/(x_0^b t^c)$ in log(L) and maximizes this log(L) over parameters b and c. Once $\tilde{b}$ and $\tilde{c}$ are obtained, $\tilde{a}$ can be computed by model 4.13. The likelihood ratio statistic for testing H_0 versus H_1 is:

$$\Lambda = -2\log L(\tilde{a}, \tilde{b}, \tilde{c}) + 2\log L(\hat{a}, \hat{b}, \hat{c}) \qquad 4.14$$

Under H_0, Λ is approximately distributed as a χ^2 distribution with a degree of freedom of 1.[41] Large values of Λ are evidence against H_0. An approximate $(1 - \alpha)\%$ confidence interval for LC($100p$) can be found by determining the set of values of x_0 for which $\Lambda \leq \chi^2(1, \alpha)$.

Confidence intervals can be based on either the approximate normal distributions of the maximum likelihood estimators, or the likelihood ratio method. A comparison of asymptotic intervals and likelihood intervals for real data sets can be found in Lawless,[35] who pointed out that, for large and moderate sample sizes, the confidence intervals obtained by these two methods are fairly close, and, for small sample sizes, likelihood ratio procedures are preferable to those based on the normal approximations. While trying to use the likelihood ratio method, we encountered a computational problem. Even to compute a single confidence interval, 20 nonlinear subroutines were needed and a large number of iterations were involved. Therefore, this method was not practical and the asymptotic normality method was relied on.

4.2.4   Data base for validation

Acute data used for analyses and chronic data used for model validation were from tests conducted at the Columbia Environmental Research Center (U.S. Geological Survey, Columbia, MO) and the National Health and Environmental Effects Research Laboratory, Gulf Ecology Division (U.S. Environmental Protection Agency, Gulf Breeze, FL) on seven fish species (rainbow trout, *Oncorhynchus mykiss;* cutthroat trout, *O. clarki;* brook trout, *Salvelinus fontinalis;* lake trout, *S. namaycush;* fathead minnow, *Pimephales promelas;* channel catfish, *Ictalurus punctatus;* and sheepshead minnow, *Cyprinodon variegatus).* With the exception of one static acute test, acute and chronic tests were conducted in flow-through diluter systems modeled after that described by Mount and Brungs.[42] Each diluter delivered four to seven concentrations of toxicant and a control. Water temperature was maintained within ±1°C of the desired temperature, and day length was regulated by the methods of Drummond and Dawson.[43] Sample sizes ranged from 10 to 30 organisms, and concentration dilution factors ranged from 50 to 75%. Acute and chronic tests were conducted in accordance with standard procedures,[21,22,44] and the chronic tests ranged from embryo-larval to full life-cycle tests. Exposure concentrations of all chemicals were measured.

The concentration-response data in the historical database for chlordane, endrin, EPN (organophosphate insecticide), heptachlor, methoxychlor, and toxaphene with sheepshead minnows were inadequate for observations preceding 96 h. For that reason, several acute tests with sheepshead minnows were conducted again. Acute and chronic flow-through tests were also conducted with carbon tetrachloride and sheepshead minnows as another test of the methods, because carbon tetrachloride is considered to have different modes of action between acute and chronic exposures with mammals.[45–47]

The acute and chronic data used in model assessment and validation can be found in Mayer et al.[29]

4.3   Validation and discussion

4.3.1   Two-step linear regression

The linear regression model regresses LCp from several acute exposure tests on the reciprocal of exposure time. The model is LCp = $a + b$ (1/t), where LCp observations are from all observation times within an acute test using the single assay methods of Finney.[48] A normal tolerance distribution is assumed in the population. As t becomes large, $1/t$ approaches zero, and the intercept is interpreted as the predicted chronic concentration, LCp. When p is small, the chronic no-effect (i.e., 0.01%) concentration for lethality is estimated. This extension[17] of the work of Bliss[25] uses univariate estimation procedures, and allows for nonparallel slopes as exposure time varies. The linear regression approach provides PNOECs for marginal data sets.

Predicted values were highly accurate for a variety of chemicals and fish species when compared with observed values for chronic tests (Table 4.1). The predicted chronic no-effect concentrations (PNOECs) were very close to, or within the limits (highest concentration without a statistically significant effect on survival and the next higher concentration with a significant effect) of the MATC for lethality and varied by less than a factor of 2 from MATCs 85% of the time. The other 15% of predictions produced values of 3.3 to 5.9 (3.3, 4.0, 4.7, 5.9) x the MATC limits. If predicted confidence limits are taken into account, the accuracy is 89% with factors greater than 2 being 2.4, 2.7 and 4.4.

The technique worked very well in predicting chronic lethality of carbon tetrachloride to sheepshead minnows (PNOEC = 8.7 mg/L, observed = $4.5 \leq x \leq 11$). Although this was only one study and one fish species, it might indicate that one assumption (i.e., the mode of action for lethality is similar under acute and chronic exposures) is not required for the technique. It may also be that carbon tetrachloride does not have different modes of action between acute and chronic exposures in fishes, as has been observed for mammals.[45-47] Although the proposed approach worked very well for several species of freshwater fish and at least one species of marine fish, its applicability to invertebrates needs to be determined.

The predictive technique was also highly accurate among various single chemicals and mixtures; it seemed representative of a wide range of octanol/water partition coefficients (log K_{ow}). Results of acute static tests might be used when acute flow-through test results are not available and the log K_{ow} is low (e.g., fluridone). Highly water soluble chemicals do not adsorb to the test container or are not taken up by the test organisms as much as chemicals of low water solubility, and exposure more closely resembles that for flow-through tests. However, additional research is needed to determine the log K_{ow} below which static test data can reliably be used to predict chronic toxicity.

The linear regression technique for deriving PNOECs uses some aspects of previously developed concepts. Sprague[20] recommended that acute tests be conducted until the toxicity curve became parallel to the time axis, indicating a threshold concentration. An incipient LC50 is then estimated by selecting an exposure time from the asymptotic part of the toxicity curve. The reciprocal of mean survival times within concentrations was used as early as 1917 by Powers.[49] Regressing the reciprocal of mean survival time on concentration to derive theoretical thresholds of toxicity was further developed by Abram[50,51] and Alderdice and Brett.[52] Although observing survival times in acute tests has merits, it is laborious and done only infrequently.

Time–toxicity relationships have also been proposed by Heming et al.[27] to study time-dependent toxicity of methoxychlor to fish. The LC50 values were related to exposure time using eight models (t = exposure time, y = LC50).

$$y = a + bt \tag{1}$$

$$y = a + b/t \tag{2}$$

$$y = a + b \ln t \tag{3}$$

$$y = 1/(a + bt) \tag{4}$$

$$y = 1/(a + b \ln t) \tag{5}$$

$$\ln y = a + bt \tag{6}$$

$$\ln y = a + b/t \tag{7}$$

$$\ln y = a + b \ln t \tag{8}$$

Models 2, 4, 7, and 8 were found to fit quite well when used on data from several fish species. Models 1, 2, and 3 are similar to the linear regression[17] and multifactor probit analysis[24] when LCp = LC50. None of the models, however, used an estimate of the concentration–time-response surface to predict chronic no-effect concentrations.

4.3.2 Multifactor probit analysis

Data from six independent acute toxicity studies that fulfilled data requirements were used for method and model selection illustrations[24]: sheepshead minnows with carbon tetrachloride, EPN, and methoxychlor; fathead minnows with kepone; brook trout with TFM and toxaphene (Table 4.1). The exposure time when the MATC was observed (duration of chronic study) was used as T_{xc} to calculate the PNOEC. Minimum (HF) and ± 2 SE values are given for PNOECs for each of the six models used on six data sets. With methoxychlor and sheepshead minnows, model B1 had the minimum heterogeneity factor among all six models used. The heterogeneity factors ranged from 0.60 for model B1 to 1.37 for model A2. Estimates of α, β, and δ after six iterations were –16.565, 6.087, and –93,949, respectively. PNOEC, as a function of exposure time, was found to be: PNOEC = exp (1.281 + 16.565 + 93.949/T)/6.087. Using T = 2,688 h (duration of chronic exposure), the model predicts PNOEC = 19 μg/L (±2 SE, 15 to 22 μg/L). The observed MATC range for lethality was 12 to 23 μg/L.

Table 4.1 Observed MATCs and Predicted Chronic No-Effect Concentrations (μg/L) and 95% CI[a] for Three Models

Chemical Species	Log K_{ow}	MATC	Linear Regression[17]	Multifactor Probit Analysis[24]	Accelerated Life Testing[30]
Butyl benzyl phthalate Fathead minnows	4.44	>360	635 (405 – 999))	880 (94 – 1379)	546 (381 – 712)
Carbon tetrachloride Sheepshead minnows	2.64	4500 – 11, 200	8734 (8084 – 9435)	8657 (6968 – 10, 017)	3423 (281 – 6564)
Chlordane Sheepshead minnows	5.8	7.1 – 17	1.5 (1.4 – 1.6)	0.19 (0.06 – 1.9)	2.0 (0.39 – 3.7)
Complex effluent Fathead minnows	–	2.0 – 3.5%	5.2 (3.8 – 6.5)%	1.9 (0.77 – 3.1)%	2.9 (2.0 – 3.9)%
2,4-D butyl ester Cutthroat trout Lake trout	2.81	24 – 44 33 – 60	147 (104 – 206) 61 (25 – 151)	209 (153 – 236) 138 (2.3 – 243)	40 (20 – 61) 42 (17 – 67)
2,4-D PGBEE Cutthroat trout Lake trout	4.88	31 – 60	96 (57 – 161) 52 – 100	ND 106 (74 – 152)	40 (33 – 46) 72 (65 – 78)
Endosulfan Sheepshead minnows	4.90-6.00[d]	0.92 – 2.1 1.1 – 2.5	2.2 (1.6 – 2.9)	0.58 (0.31 – 0.86)	0.53 (0.11 – 0.95)
Endrin Sheepshead minnows	4.56-5.30	0.12 – 0.31	0.03 (0.002 – 0.42)	0.007 (0.004 – 0.011)	0.007 (0.001 – 0.014)
EPN Sheepshead minnows	4.8	4.1 – 7.9	3.8 (0.15 – 92)	2.0 (0.87 – 4.4)	3.3 (0.66 – 6.0)
Fluridone Channel catfish[b]	1.87	1000 – 2000	1183 (–1079 – 3444)	243 (14 – 1065)	250 (50 – 449)
Heptachlor Sheepshead minnows	5.44	1.9 – 2.8 2.2 – 3.5	2.6 (0.37 – 4.9)	2.4 (1.6 – 3.2)	1.9 (0.20 – 3.6)

Chemical / Species					
Kepone	6.08				
Fathead minnows		1.2 – 3.1	2.9 (0.69 – 5.0)	1.2 (0.0 – 4.1)	1.9 (0.37 – 3.3)
Methoxychlor	4.2				
Brook trout[c]		1.1 – 3.1	0.82 (0.012 – 55)	0.09 (0.03 – 0.21)	0.55 (0.11 – 0.98)
Sheepshead minnows		12-23	12 (0.94 – 24)	7.9 (0 – 18)	2.3 (1.4 – 3.1)
		23 – 48	12 (9.8 – 14)	16 (11 – 18)	3.8 (1.5 – 6.1)
			18 (14 – 22)	19 (15 – 22)	5.7 (3.4 – 8.0)
Pentachlorophenol	5.01				
Fathead minnows		>142	240 (238 – 243)	ND	301 (251 – 351)
Phorate	3.5				
Sheepshead minnows		0.24 – 0.41	0.15 (0.05 – 0.41)[c]	0.17 (0.10 – 0.27)	0.21 (0.05 – 0.37)
Pydraul 50E	4.62-6.08[e]				
Fathead minnows		317 – 752	592 (296 – 888)	470 (377 – 555)	193 (151 – 235)
TFM	—				
Brook trout (adult)		4000 – 8000	4311 (3319 – 5303)	3014 (2367 – 3535)	884 (284 – 1485)
Brook trout (juvenile)		940 – 1600			
Toxaphene	4.83				
Brook trout (adult)		0.14 – 0.29	1.7 (0.79 – 3.7)	0.14 (0.05 – 0.31)	0.17 (0.02 – 0.32)
Brook trout (juvenile)		0.07 – 0.14	0.04 (-0.57 – 0.65)	0.11 (0.0 – 0.44)	0.11 (0.07 – 0.14)
Fathead minnows		0.62 – 1.3	1.8 (1.1 – 2.9)	2.7 (1.3 – 3.9)	0.32 (0.025 – 0.61)
Channel catfish		0.07 – 0.13	0.06 (-0.33 – 0.44)	0.07 (0.05 – 0.11)	0.08 (0.02 – 0.15)
Sheepshead minnows		1.1 – 2.5	0.77 (0.51 – 1.2)	0.5 (0.37 – 0.63)	0.3 (0.15 – 0.44)

a ± 2 SE for linear regression and multifactor probit analysis.

b Based on static test.

c Acute toxicity test for rainbow trout was unavailable and predicted values are based on brook trout because of similarity in response to toxicants.[69]

d Endosulfan I = 4.90, endosulfan II = 6.00.

e Pydraul 50E is a hydraulic fluid consisting of three components: triphenyl phosphate = 4.62, nonylphenyl diphenyl phosphate = 5.93, cumylphenyl diphenyl phosphate = 6.08.

An HF less than 1.0 was found for at least one of the models tried with each chemical except kepone, where goodness-of-data fit was less than that for the other five data sets. Ranges in HFs among models within an experiment were: carbon tetrachloride (range = 0.1), EPN (0.44), kepone (0.21), methoxychlor (0.77), TFM (0.84), and toxaphene (2.2). Small HFs existed for models B1 through C2 (methoxychlor); B2 through C2 (TFM), and B2 and C2 (toxaphene).

Model choices for each experiment were made using minimum HF (minHF) as a criterion, and the model with smallest uncertainty (±2 SE) was also identified. Use of the minHF to select the model resulted in PNOECs that compared favorably with MATCs for carbon tetrachloride, kepone, methoxychlor, and toxaphene. The model choice for EPN had an MATC (4.1 to 7.9 µg/L) within the ±2 SE interval estimate (0.87 to 4.4), but the PNOEC (2.0) was half the lower limit of the MATC. For TFM, PNOECs for all models were below the MATC with the model choice for PNOEC being 3,014 µg/L (MATC = 4,000 to 8,800 µg/L).

Models chosen for PNOECs based on the minimum interval of ±2 SE had PNOECs within the MATC interval for both carbon tetrachloride and toxaphene. However, with all other chemicals, this criterion identified models yielding PNOECs that were not within the MATC intervals. This suggests that models chosen on the basis of minHF, which identifies the model with the best data fit, have the best point estimates; however, the SE interval estimates are not always the smallest. Models selected for PNOECs based on minimum SE intervals may result in biased estimates of PNOECs.

Data from 22 other studies[29] with marginal data for multifactor probit analysis, were analyzed to compare PNOECs for lethality with reported MATC values for lethality (Table 4.1). Data for three acute tests (2,4-D PGBEE, cutthroat and lake trout; pentachlorophenol, fathead minnows) failed to converge for any of the models tried. This was caused by an insufficient number of concentrations or partial mortalities. However, PNOECs were determined for these three chemicals by the method of Mayer et al.[17] and Sun et al.[30] Ten of the remaining 25 data sets provided estimates for at least some of the models and had point estimates of PNOECs within the range of the reported MATCs. Eighteen of those 25 data sets had MATCs that fell within ±2 SE of the respective PNOECs. The seven remaining data sets had MATCs that were outside the range of PNOEC (±2 SE). Overall, the PNOECs were within a factor of two of the MATC limits 59% of the time. The other 41% of predictions produced values of 2.1, 2.2, 2.3, 6.0, 4.0, 4.8, 12, 4.1, 4.8 and 37 times the MATC limits or could not be calculated.If predicted confidence limits are considered, accuracy is 74%, with factors greater than 2 being 3.5, 5.2, and 11.

Multifactor probit analysis is applicable to a wide array of chemical classes and chronic test types, ranging from embryo-larval to full life-cycle tests. The models utilize several response patterns over time and are, to a slight degree, similar to those used by other researchers, 53, 54 but estimate different values.

4.3.3   Accelerated life testing

This model[30] uses natural log concentration to normalize toxicity data; data normalization is done by the ACE software program.[55] The procedure calculates the LC (r) at "acceptable levels" of 0.1 %, 0.5%, 1%, etc., for any time and concentration combination.

For calculating predicted values, 1% was used as an acceptable risk level. The ability to determine significant differences in organism populations usually diminishes below 1%, and the accelerated life-testing model is based on population responses. Smaller percentages, such as 0.01% or 0.1%, will produce smaller predicted values and probably cause incomparability between the predicted values and the MATC. However, in situations other than comparison of predicted no-effect concentrations with the NOEL or MATC, 0.01% or 0.1% could be used. In fact, a negligible risk level (i.e., 0.01% or zero effect; Mayer et al.[17]) of a chemical can be determined and probably represents true no effect, but this has not been validated for accelerated life testing.

The no-effect concentrations predicted by this technique compare well, in most cases, with the no-effect concentrations from chronic toxicity experiments for a database of a variety of chemicals and fish species (Table 4.1, Figure 4.4). When the predicted values were compared with the no-observed-effect concentrations in chronic tests, the technique provided accurate predictions. The confidence intervals for predicted no-effect concentrations and the limits of the maximum acceptable toxicant concentration overlapped or were within a factor of 2 74% of the time. The other 26% (brook

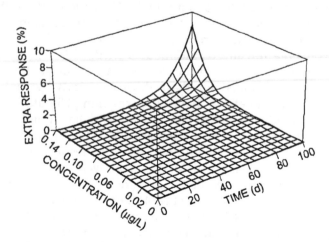

Figure 4.4 Dose–time-response surface for toxaphene juvenile-brook trout acute data[71] extended in time. Reprinted with permission from K. Sun, G.F. Krause, F.L. Mayer, M.R. Ellersieck, and A.P. Basu. Predicting chronic lethality of chemicals to fishes from acute toxicity test data: Theory of accelerated life testing. *Environmental Toxicology and Chemistry*, Vol. 14(10) 1745–1752. Copyright Society of Environmental Toxicology and Chemistry (SETAC), Pensacola, FL, 2001.

trout —TFM and toxaphene; sheepshead minnows — chlordane, endrin, fluridone, methoxychlor, and toxaphene) of predicted values for no-effect concentrations varied by factors of 3.2 to 17 (3.2, 3.6, 3.7, 4.0, 4.5, 5.2, 17) from the MATC limits. However, the accelerated life-testing model and the MATC are based on two different concepts or derivations, and hence, the predicted value from the model may not be comparable with MATCs obtained from statistical hypotheses tests.

Our predicted results were found to be lower than the MATC with seven data sets (chlordane, endrin, fluridone, methoxychlor, and toxaphene with sheepshead minnows; floridone with channel catfish; TFM and toxaphene with brook trout), and conflicting results could be caused by the uncertainty of the MATC. Specifically, it relates to the oversimplistic but widely used practice of interpreting a concentration that is not statistically different from the control as the NOEL. Such a practice ignores the concept of statistical power in hypothesis testing and often overestimates the true no-effect level.

It is common practice that a NOEL or MATC statement is made based on a significance test from several dose-response observations. Hence, a good interpretation of the NOEL depends on a thorough understanding of hypothesis testing and statistical significance. The accuracy of a statistical-testing conclusion depends on type I error, type II error, sample size, design, and the variability in response.[56] Type II error or power of a test (1 minus type II error) is of importance for assessing the discriminating ability of the decision rule of a statistical hypothesis test. The smaller and less precise the experiment, the larger the type II error, and hence, the smaller the power of detecting a dose effect. Constrained by time and cost, most practical toxicity tests have only moderate sample size and precision; thus, they incur a substantial type II error rate. In fact, it is not unusual that a statistical hypothesis test has a power of 60 to 80%. Therefore, the MATC approach has quite limited ability to detect true no-effect concentrations (i.e., MATCs were found to be in conflict with acute toxicity test observations in seven data sets).

In the acute endrin-sheepshead minnow test, 55% mortality was observed at 0.26 µg/L and 3 d of exposure. This indicates that the estimate for the no-effect concentration at 3 d should be lower than 0.26 µg/L because the no-effect concentration causes negligible mortality. When considering a long duration of exposure, the estimated no-effect concentration should be even lower than that for an acute exposure. However, the MATC was 0.12 to 0.31 µg/L for 140 d of exposure. Based on the observations from the acute test, it is reasonable to believe that this chronic result overestimates the no-effect concentration. The same situation occurs with the other six data sets except for 3-trifluoromethyl-4-nitrophenol (TFM). The acute and chronic data used for deriving the chronic LC01 and MATC were based on adult fish for TFM. The MATC for fry produced by the adults was 940 to 1,600 µg/L, which compares well with the 885 µg/L chronic LC01 predicted with accelerated life testing.

Low power for detecting concentration effects may be the statistical reason for the appearance of any conflicting conclusions. As Hoekstra and Van Ewijk[15] stated, the NOEL approach often leads to higher values for the no-effect levels. In contrast, our procedure, using an improved method and making a more complete use of the experimental information, predicted exposure effects at concentrations below those estimated by the MATC approach. Also, iterative numerical techniques produce confidence limits not available for the MATC approach.

The accelerated life testing model and procedure require three-dimensional data: time–concentration response. Quantal-response data at one point in time, of course, cannot be used to predict chronic lethality, as this kind of data does not include a time factor. Like the multifactor probit method, there is also a restriction on the data that can be used in the proposed model. For the model to function, at least three partial mortalities are required in the total acute data set, and it is obvious that with more partial mortalities one can obtain more accurate estimates of the parameters.

In summary, this analysis is the statistical procedure for observations on time to death from Design II. This is the most common experiment conducted, and this analysis does not require independent fraction dead at time *t* associated with each concentration. This method (Design II and based on accelerated life testing) is therefore superior to alternatives that use Design II and procedures based on regression analysis.

4.3.4   General

Intercomparisons of acute and chronic toxicity have been reviewed by Giesy and Graney.[57] Other aspects of predictions with acute toxicity data have been presented by Breck[53] (fluctuating concentrations), Mancini[54] (time variable concentrations of toxicants), Dixon and Newman[9] and Newman et al.[3,10,34] (survival time modeling). An alternative for no-observed-effect levels was developed by using linear extrapolation to derive "acceptable" small-effect concentrations (>1 ≤ 25%).[15]

Another use of acute toxicity data to estimate chronic lethality of a chemical is the toxicity threshold value or LC1,[58,59] which is calculated for 1.0% mortality and at one point in time. This application of acute tests should work well for those chemicals, effluents, etc. that differ little in toxicity between acute and chronic effects or where the LC1 is derived at a duration approaching or within chronic exposure conditions. However, the LC1 does not take into account time course of effect, and its use for predictive purposes is limited for a wide range of chemicals, particularly those that bioconcentrate significantly or have cumulative effects. A comparison of the PNOECs to LC1s (ratio of PNOEC of LC1 to the geometric mean of the lower and upper MATC limits) resulted in means of 1.2 (PNOEC) and 4.2 (LC1) with ranges of 0.19 to 8.5 and 0.48 to 33, respectively (i.e., LC1s averaged 3.5 times higher than PNOECs and 4.2 times higher than geometric mean MATCs).

During the last 30 years, the most commonly conducted fish chronic toxicity tests have been shortened from full-life-cycle tests to 30- to 90-d early-life-stage or partial-life-cycle tests[60,61] and then to 7-d subchronic tests.[62] Review of subchronic, early-life-stage, partial-life-cycle, and full-life-cycle toxicity tests with several fish species demonstrated that the shorter tests are good estimators of chronic toxicity (lethality and growth) and of MATCs observed in the longer life-cycle tests.[60-65] Although the success of developing more rapid tests to estimate chronic toxicity is empirically based, it does corroborate the toxicological concept of time course of effect in using acute data to predict chronic lethality.

A shortcoming of the MATC is that it is based on just a single endpoint (mortality, growth or reproduction), reflecting the exposure concentration at the termination of a toxicity test, and the time course of effect is usually ignored. To make sound and informed estimates of the chronic no-effect concentration of a chemical, it is necessary to understand that the toxic impact of a chemical to aquatic organisms in a homogeneous population is a progressive development through time of exposure. Not only data at one point in time, but also the observed responses at other times in the experiment should be used. Our procedures utilize all acute test data and employ more advanced statistical theory and technique. Hence, these methods increase the ability to detect toxic effects of chemicals and reduce uncertainty in assessment of chronicity.

One limitation of these models is that only chronic mortality of a chemical is analyzed and no other endpoints of chronic toxicity (i.e., growth and reproduction). Therefore, further work is needed to investigate the statistical relationships among other endpoints with mortality, such as egg hatchability and organism growth, and to develop theory and methodology to describe this relationship and make reliable inferences. Some preliminary work in this area is found in Mayer et al.[66] and Suter et al.[67] Assessments of sensitivity in relation to chronic endpoints in fishes[66-68] indicated that survival and growth are often equally sensitive, and growth may not be of critical importance in establishing NOECs in most tests. In tests for which growth was the single most sensitive endpoint, survival could be used to estimate the NOEC for growth within a factor of three.[66]

NOECs are nearly always lower for reproduction endpoints than for survival.[66,67] Attempts to relate acute mortality to chronic reproductive effects by regression analysis have not been successful.[67] Because of the likelihood of different modes of action between lethal and reproductive effects, we do not recommend that reproductive effects be predicted using the proposed method. Even so, the techniques presented are highly beneficial in the preliminary assessment of chronic toxicity of chemicals and effluents and in predicting chronic no-effect concentrations for survival and growth with species that are difficult to culture under chronic testing conditions. Finally, chronic toxicity predictions from well-conducted acute tests appear very adequate.

4.4   Conclusions

Because chronic mortality is an endpoint that is consistent with acute test results and is easier to observe, the concepts and models developed for predicting chronic lethality of chemicals herein should provide a framework that enables transfer of this knowledge to other aspects of chronicity. Additional assumptions may be required. One is chemical toxicity by a single mode of action and by cumulative effects over time and concentration. Another is that the biological mechanism for tolerance or lethality at low concentrations is the same as at high concentrations, validating extrapolation to low concentrations. The third assumption is that the model fits the data. Goodness-of-fit and other (e.g., residual plot) tests can be used to check if the correct distribution is chosen.

The advantage of the linear regression technique is that it will generally provide estimates when the multifactor probit analysis and accelerated life testing models will not; partial mortalities are not required. However, it is limited to only one point in time, and that is infinity (intercept of regression). Linear regression is the most accurate (> 85–89% of the time, ≤ 2 x MATC) when estimating no-effect chronic concentrations based on hypothesis testing.

Multifactor probit analysis uses the iterative reweighted least-squares method to estimate the parameters of the probit surface. Independent variables are time of exposure and concentration of toxicant; the dependent variable is probit proportion responding. This model allows prediction of the concentration of a toxicant at any time and percentage mortality, including the 96-h LC50. However, at least five partial mortalities are recommended for proper functioning. Multifactor probit analysis is a special case model and is recommended for experimental designs where each time interval is completely independent of any other (e.g., removing all organisms at 24 h and using different exposure containers and organisms for 48 h, etc.). This model was the least accurate (59–74% of the time, ≤ 2 x MATC) for predicting chronic no-effect concentrations established with hypothesis testing.

Accelerated life testing is a statistical procedure for observations on time to death and is technically superior to linear regression and multifactor probit analysis. Three partial mortalities are recommended, but it will function with one. The model allows input of any chronic time in question for low proportions of effect (i.e., 0.01, 0.05, 0.1, 0.5, 1.0, and 5.0%). Accelerated life testing was intermediate in accuracy (74–81% of the time, ≤ 2 x MATC) for predicting chronic no-effect concentrations. However, this is probably due to the low power of hypothesis testing in detecting concentration effects, often leading to higher values for no-effect concentrations. This model should be capable of predicting zero population effect, but support of that hypothesis will require validation in chronic tests exposing very large numbers of organisms.

The accelerated life testing and linear regression procedures are available as software,[55] as well as multifactor probit analysis.[29] These three models are being combined into a Windows-compatible version.

Acknowledgments

This project was sponsored in part by the U.s. Environmental Protection Agency's Offices of Research and Development, Pesticide Programs, Pollution Prevention and Toxics, and Water. We also acknowledge V.A. Coseo for typing the manuscript.

References

1. Chen, F. and Schüürmann, G., *Quantitative Structure-Activity Relationships in Environmental Sciences*. VII, SETAC Press, Pensacola, FL, 1997.

2. Finney, D.J., *Probit Analysis*, 3rd ed., Cambridge University Press, London, UK, 1971.

3. Newman, M.C., *Quantitative Methods in Aquatic Ecotoxicology*, Lewis Publishers, Boca Raton, FL, 1994.

4. Mount, D.I. and Stephan, C.E., A method for establishing acceptable limits for fish – malathion and the butoxyethanol ester of 2,4-D, *Transactions of the American Fisheries Society*, 96, 185, 1967.

5. Kenaga, E.E., Predictability of chronic toxicity from acute toxicity of chemicals in fish and aquatic invertebrates, *Environ. Toxicol. & Chem.*, 1, 347, 1982.

6. Kenaga, E.E., Aquatic test organisms and methods useful for assessment of chronic toxicity of chemicals, in *Analyzing the Hazard Evaluation Process*, Dickson, K.L., Maki, A.W. and Cairns, J., Jr., Eds., American Fisheries Society, Washington, D.C., 1979, 101.

7. Buikema, A.L., Jr., Niederlehner, B.R. and Cairns, J., Jr., Biological monitoring. 4. Toxicity testing, *Water Research*, 16, 239, 1982.

8. Shirazi, M.A. and Lowrie, L., Comparative toxicity based on similar asymptotic endpoints, *Arch. Environ. Contam. & Toxicol.*, 17, 273, 1988.

9. Dixon, P.M. and Newman, M.C., Analyzing toxicity data using statistical models for time-to-death: an introduction, in *Metal Ecotoxicology: Concepts and Applications*, Newman, M.C. and Mcintosh, A.W., Eds., Lewis, Chelsea, MI, 1991, 207.

10. Newman, M.C. and Aplin, M.S., Enhancing toxicity data interpretation and prediction of ecological risk with survival time modeling: an illustration using sodium chloride toxicity to mosquitofish, *Aquatic Toxicology*, 23, 85, 1992.

11. Gelber, R.D., Lavin, P.T., Mehta, C.R., and Schoenfeld, D.A. Statistical analysis of toxicity testing, in *Fundamentals of Aquatic Toxicology*, Rand, G.M. and Petrocelli, S.R., Eds., Hemisphere, New York, NY, 1985, 110.

12. Crump, K.S., A new method for determining allowable daily intakes, *Fundam. & Appl. Toxicol.*, 4, 854, 1984.

13. Gaylor, D.W., Quantitative risk analysis for quantal reproductive and developmental effects, *Environ. Hlth. Perspect.*, 79, 243, 1989.

14. Chen, J.J. and Kodell, R.L., Quantitative risk assessment for teratological effects, *J. Am. Stat. Assoc.*, 84, 966, 1989.

15. Hoeskstra, J.A. and Van Ewijk, P.H., Alternative for the no-observed-effect level, *Environ. Toxicol. & Chem.*, 12, 187, 1993.

16. Hartley, H.O. and Sielken, R.L., Jr., Estimation of "safe doses" in carcinogenic experiments, *Biometrics*, 33, 1, 1977.

17. Mayer, F.L., Krause, G.F., Buckler, D.R., Ellersieck, M.R. and Lee, G., Predicting chronic lethality of chemicals to fishes from acute toxicity test data: concepts and linear regression, *Environ. Toxicol. & Chem.*, 13, 671, 1994.

18. Litchfield, J.T., Jr. and Wilcoxon, F., A simplified method of evaluating dose-effect experiments, *J. Pharmacol. Exper. Ther.*, 96, 99, 1949.

19. Green, R.H., Estimation of tolerance over an indefinite time period, *Ecology*, 46, 887, 1965.

20. Sprague, J.B., Measurement of pollutant toxicity to fish. 1. Biossay methods for acute toxicity, *Water Research*, 3, 793, 1969.

21. American Society for Testing and Materials, Standard practice for conducting acute toxicity tests with fishes, macro-invertebrates, and amphibians, in *Annual Book of ASTM Standards, E729-80*, Philadelphia, PA, 1980, 1.

22. Committee on Methods for Toxicity Tests with Aquatic Organisms, *Methods for Acute Toxicity Tests with Fish, Macroinvertebrates, and Amphibians*, EPA 660/3-75-009, Ecological Research Series, United States Environmental Protection Agency, Corvallis, OR, 1975.

23. Snedecor, G.W. and Cochran, W.G., *Statistical Methods*, 7th ed., Iowa State University Press, Ames, IA, 1980.

24. Lee, G., Ellersieck, M.R., Krause, G.F. and Mayer, F.L., Predicting chronic toxicity of chemicals to fishes from acute toxicity test data: multifactor probit analysis, *Environ. Toxicol. & Chem.*, 14, 345, 1995.

25. Bliss, C.I., The relation between exposure time, concentration and toxicity in experiments on insecticides, *Ann. Entomol. Soc. Amer.*, 33, 721, 1940.

26. Finney, D.J., Examples of the planning and interpretation of toxicity tests involving more than one factor, *Ann. Appl. Biol.*, 29, 330, 1942.

27. Heming, T.A., Arvind, S. and Yogesh, K., Time-toxicity relationships in fish exposed to the organochlorine pesticide methoxychlor, Environmental Toxicology and Chemistry, 3, 923, 1989.

28. Finney, D.J., *Statistical Methods in Biological Assay*, Charles Griffin, London, UK, 1978.

29. Mayer, F.L., Krause, G.E., Ellersieck, M.R. and Lee, G., *Statistical Approach to Predicting Chronic Toxicity of Chemicals to Fishes from Acute Toxicity Test Data*, PB92-169655; PB92503119 (software), National Technical Information Service, Springfield, VA, 1991.

30. Sun, K., Krause, G.F., Mayer, F.L., Ellersieck, M.R. and Basu, A.P., Predicting chronic lethality based on the theory of accelerated life testing, *Environ. Toxicol. & Chem.*, 14, 1745, 1995.

31. Kalbfleisch, J.D. and Prentice, R.L., *The Statistical Analysis of Failure Time Data*, John Wiley & Sons, New York, NY, 1980.

32. Cox, D.R., Regression models and life-tables with discussion, J. Royal Stat. Soc., B, 34, 87, 1972.

33. Armitage, P. and Doll, R., Stochastic models for carcinogenesis, in *Proceedings of the Fourth Berkeley Symposium on Mathematical Statistics.and Probability*, Vol. 4, Neyman, J., Ed., University of California Press, Berkeley, CA, 1961, 19.

34. Newman, M.C., Keklak, M.M. and Doggett, M.S., Quantifying animal size effects on toxicity: a general approach, *Aquat. Toxicol.*, 28, 1, 1994.

35. Lawless, J.F., *Statistical Models and Methods for Lifetime Data*, John Wiley & Sons, New York, NY, 1982.

36. Shirazi, M.A. and Lowrie, L., A probabilistic statement of the structure activity relationship for environmental risk analysis, *Arch. Environ. Contam. & Toxicol.*, 19, 597, 1990.
37. Agresti, A., *Categorical Data Analysis*, John Wiley & Sons, New York, NY, 1990.
38. Cox, D.R. and Hinkley, D.V., A note on the efficiency of least squares estimates, *J. Royal Stat. Soc.*B, 30, 284, 1968.
39. Nelson, W.B., *Applied Life Data Analysis*, John Wiley & Sons, New York, NY, 1982.
40. Rao, C.R., *Linear Statistical Inference and Its Applications*, John Wiley & Sons, New York, NY, 1965.
41. Wilks, S.S., The large-sample distribution of the likelihood ratio for testing composite hypotheses, *Ann. Math. & Stats.*, 9, 60, 1938.
42. Mount, D.I. and Brungs, W.A., A simplified dosing apparatus for toxicology studies, *Water Research*, 1, 21, 1967.
43. Drummond, R.A. and Dawson, W.F., An inexpensive method for simulating diel patterns of lighting in the laboratory, *Trans. Amer. Fish. Soc.* 99, 434, 1970.
44. American Public Health Association, American Water Works Association and Water Pollution Control Federation, *Standard Methods for the Examination of Water and Wastewater*, 17th ed. American Public Health Association, Washington, D.C., 1989.
45. Haley, T.J., Solvents and chemical intermediates, in *Handbook of Toxicology*, Haley, T.J. and Berndt, W.O, Eds., Hemisphere, Washington, D.C., 1987, 504.
46. Hardin, B.L., Carbon tetrachloride poisoning — a review, *Indust. Med. & Surg.*, 23, 93, 1954.
47. Recknagel, R.O., Glende, E.A., Jr., Dolak, J.A. and Waller, R.L., Mechanisms of carbon tetrachloride toxicity, *Pharmacol. Ther.* 43, 139, 1989.
48. Finney, D.J., The statistical treatment of toxicological data relating to more than one dosage factor, *Ann. Appl. Biol.*, 30, 71, 1943.
49. Powers, E.B., The goldfish (*Carassius carassius*) as a test animal in the study of toxicity, *Illinois Biol. Mono.*, 4, 127, 1917.
50. Abram, F.S.H., An application of harmonics to fish toxicology, *Int. J. Air & Water Poll.*, 8, 325, 1964.
51. Abram, F.S.H., The definition and measurement of fish toxicity thresholds, *Proceedings, Third International Conference, Advances in Water Pollution Research*, Munich, West Germany, September 1966, 75, 1967.
52. Alderdice, D.F. and Brett, J.R., Some effects of kraft mill effluent on young Pacific salmon, *J. Fish. Res. Bd. Can.*, 14, 783, 1957.
53. Breck, J.E., Relationships among models for acute toxic effects: applications to fluctuating concentrations, *Environ. Toxicol. & Chem.*, 7, 775, 1988.
54. Mancini, J.L., A method for calculating effects on aquatic organisms, of time-varying concentrations, *Water Research* 17, 1355, 1983.
55. Mayer, F.L., Sun, K., Lee, G., Ellersieck, M.R. and Krause, G.F., *User Guide: Acute to Chronic Estimation (ACE)*, EPA/600/R-98/152, United States Environmental Protection Agency, Gulf Breeze, FL, 1999.
56. Neter, J., Wasserman, W. and Kutner, M.H., *Applied Linear Statistical Models*, 2nd ed., Irwin, Homewood, IL, 1985.
57. Giesy, J.P. and Graney, R.L., Recent developments in and intercomparisons of acute and chronic bioassays and bioindicators, *Hydrobiologia*, 188/189, 21, 1989.

58. Birge, W.P., Black, J.A. and Westerman, A.G., Short-term fish and amphibian embryo-larval tests for determining the effects of toxicant stress on early life stages and estimating chronic values for single compounds and complex effluents, *Environ. Toxicol. & Chem.*, 4, 807, 1985.

59. Birge, W.P., Black, J.A., Short, T.M. and Westerman, A.G., A comparative ecological and toxicological investigation of a secondary wastewater treatment plant effluent and its receiving stream, *Environ. Toxicol. & Chem.*, 8, 437, 1989.

60. McKim, J.M., Evaluation of tests with early life stages of fish for predicting long-term toxicity, *J. Fish. Res. Bd. Can.*, 34, 1148, 1977.

61. McKim, J.M., Early life stage toxicity tests, in *Fundamental of Aquatic Toxicology*, Rand, G.M. and Petrocelli, S.R., Eds., Hemisphere, Washington, D.C., 1985, 58.

62. Norberg, T.J. and Mount, D.I., A new fathead minnow (*Pimephales promelas*) subchronic toxicity test, *Environ. Toxicol. & Chem.*, 4, 711, 1985.

63. Macek, K.J. and Sleight, B.N., III., Utility of toxicity tests with embryos and fry of fish in evaluating hazards associated with the chronic toxicity of chemicals to fishes, in *Aquatic Toxicology and Hazard Evaluation* (First Symposium), Mayer, F.L. and Hamelink, J.L., Eds., STP 634, American Society for Testing and Materials, Philadelphia, PA, 1977, 137.

64. Norberg-King, T.J., An evaluation of the fathead minnow seven-day subchronic test for estimating chronic toxicity, *Environ. Toxicol. & Chem.*, 8, 1075, 1989.

65. Woltering, D.R., The growth response in fish chronic and early life stage toxicity tests: a critical review, *Aquat. Toxicol.*, 5, 1, 1984.

66. Mayer, F.L., Mayer, K.S. and Ellersieck, M.R., Relationship of survival to other endpoints in chronic toxicity tests with fish, *Environ. Toxicol. & Chem.*, 5, 737, 1986.

67. Suter, G.W., II, Rosen, A.E., Linder, E. and Parkhurst, D.F., Endpoints for responses of fish to chronic toxic exposures, *Environ. Toxicol. & Chem.*, 6, 793, 1987.

68. Ward, G.S. and Parrish, P.R., Evaluation of early lifestage toxicity tests with embryos and juveniles of sheepshead minnows (*Cyprinodon variegatus*), in, *Aquat. Toxicol.* (Third Symposium), Eaton, J.G, Parrish, P.R. and Hendricks, A.C., Eds., STP 707, American Society for Testing and Materials, Philadelphia, PA, 1980, 243.

69. Mayer, F.L., Deans, C.H. and Smith, A.G., *Inter-taxa Correlations for Toxicity to Aquatic Organisms*, EPA600/8-87-017, United States Environmental Protection Agency, Gulf Breeze, FL. 1987.

70. Buckler, D.R., Witt, A., Jr., Mayer, F.L., and Huckins, J.N., Acute and chronic effects of Kepone and mirex on the fathead minnow, *Trans. Amer. Fish. Soc.*, 110, 270, 1981.

71. Mayer, F.L., Mehrle, P.M. and Dwyer, W.P., *Toxaphene Effects on Reproduction, Growth, and Mortality of Brook Trout*, EPA 600/3-75-013, Ecological Research Series, United States Environmental Protection Agency, Duluth, MN, 1975.

chapter 5

Just how much better is a time to event analysis?

Philip M. Dixon

Contents

5.1 Introduction

In a traditional fixed-time study of toxicant effects, the number of alive and dead individuals in each experimental condition is recorded at a fixed time (e.g., 96 h). If the purpose is to estimate LC50, the median lethal concentration, the experimental conditions are different concentrations. If the purpose is to evaluate individual or environmental characteristics that affect sensitivity to a toxicant, the experimental conditions are groups with different individual characteristics or different exposure environments. Estimates of LC50 or treatment effects are based on the proportion of dead individuals in each experimental condition. Many different techniques, including probit analysis, logistic regression, the trimmed Spearman-Karber approach, and contingency table analysis can be used to estimate dose-response curves, LC50s, or treated effects. These techniques are described in many textbooks;[1] Hamilton[2] reviews the statistical properties of these methods.

In a time to death study, more information about the death time is recorded for each individual. Ideally, the exact time of death of an individual is recorded. Intuitively, it seems obvious that recording exact death times should provide more information about dose-response curves and treatment effects.[3] However, the magnitude of the improvement and how to measure the improvement are unclear.

Practical experimental constraints usually introduce right censoring. Some individuals are likely to have very long times to death, either because they are quite robust or they are in benign experimental conditions. Continuing the experiment long enough to record the last few death times is inefficient. Instead, the experiment can be terminated at an arbitrary time, T_c. Individuals alive at the end of the experiment are right censored ($t > T_c$); the only information available about their time of death is that it is longer than the length of the experiment. Many different statistical methods can be used to estimate LC50, the dose-response curve, or treatment effects from censored time to event data. The choices include nonparametric techniques that make few assumptions about the data, semiparametric techniques that use a specific model for dose-response or differences among groups, and fully parametric models that specify all aspects of the distribution of death times.[4] Censoring reduces the information in a time to death data set, and it is no longer clear that a time to event study provides more precise estimates.

This chapter considers whether time to event methods for censored data provide better estimates of one very important toxicological parameter, LC50. First, it describes the two models, defines some statistical aspects of "better" and justifies a focus on the precision of an estimate. The author derives a way to estimate LC50 and its precision from time to event data. These procedures are illustrated using data from two classic studies. The estimates using the time to event model are more precise, but the improvement in precision is quite inconsistent. Fisher Information[5] is used to quantify the amount of information provided by observations from fixed-time data and time to death data and identify when time to death data lead to more precise estimates.

5.2   Comparing traditional and time to event models

In a parametric time to event model the distribution of death times is described by a small number of parameters. Consider modeling data with a linear dose-response curve and a log-logistic distribution of times to death. This model can be written either as a regression-like model for log-transformed time to death or as a model for the fraction of surviving individuals. If T_i is the observed time to death for an individual exposed to concentration C_i, α and β are the intercept and slope of the dose-response curve, and γ is the variability parameter for the log-logistic distribution, then

$$\log T_{ij} = \mu_i + e_{ij} = \alpha + \beta C_i + e_{ij}, \; e_{ij} \sim \text{Logistic } (0, \gamma) \qquad 5.1$$

This model describes how the mean log-transformed time to death, μ_i, is influenced by the exposure concentration ($\mu_i = \alpha + \beta C_i$) and how individual observations vary around that mean ($\log T_{i,j} \sim \text{Logistic } (\mu_i, \gamma)$).

Alternatively, the fraction of individuals that die before time t in the group exposed to concentration C_i is given by the cumulative distribution function

$$F(t) = \frac{1}{(1 + e^{\alpha + \beta C_i - \gamma \log t})} \qquad 5.2$$

and the fraction of individuals alive at time t is given by the survivor function

$$S(t) = \frac{1}{(1 + e^{\gamma \log t - \alpha - \beta C_i})} \qquad 5.3$$

These models for survival (Equation 5.3) or death (Equation 5.2) can be linearized using a logit transformation.

$$\log\left(\frac{S(t)}{1 - S(t)}\right) = -\log\left(\frac{F(t)}{1 - F(t)}\right) = \alpha + \beta C_i - \gamma \log t \qquad 5.4$$

The assumptions of linearity and constant scale in Equations 5.1 and 5.4 can be relaxed, at the cost of complicating the analysis. Given a model (e.g. Equation 5.1), the unknown parameters and their asymptotic variances can be estimated by maximum likelihood. Hypotheses can be tested by likelihood ratio tests. Dixon and Newman[4] describe these techniques in more detail and give examples of their use for toxicological questions and data.

The log-logistic time to death model (Equation 5.4) is a generalization of a linear model for the proportion of surviving animals in a traditional fixed-time experiment.

$$\text{logit}(p_i) = \log\left(\frac{p_i}{1 - p_i}\right) = a + b C_i \qquad 5.5$$

where p_i is the fraction of individuals surviving to some fixed time at dose C_i. Although Equation 5.4 is better known with a probit function in place of the logit function, the differences between the two functions are small, except for proportions very close to zero or one.[3] If the log-logistic distribution for times to death were replaced by a lognormal distribution, the corresponding linearized model for survival time (analogous to Equation 5.4) would be a probit model. The linear logit model, Equation 5.5, can be reparameterized as a non-linear logit model with LC50 as one of the parameters.

$$\text{logit}(p_i) = \log\left(\frac{p_i}{1 - p_i}\right) = b(C_i - \text{LC50}) \qquad\qquad 5.6$$

The traditional dichotomy of alive or dead at a fixed time (e.g., 96 h) is a specific type of censoring in a time to event analysis. Live animals are right-censored individuals (time to death > 96 h); dead animals are left-censored (time to death < 96 h). In time to death data, the left-censored data are replaced by the observed death times. Hence, both traditional and time to event methods are estimating similar quantities. Given two methods for estimating the same quantities, is one better than another?

Statistical procedures can be evaluated by multiple criteria, including toxicological objectives (e.g., the goals of the analysis), statistical precision, statistical assumptions, and simplicity (Table 5.1). Time to event methods may answer a broader range of questions because they explicitly include time and concentration (or other experimental conditions) in the analysis and they can include individual characteristics (such as body size or genetic information). Other chapters in this book illustrate the range of questions that can be answered with time to event methods in toxicology and other disciplines. Time to event methods have also been used to extrapolate data from acute toxicity trials to estimate mortality in long-term low-level exposures (e.g., Ref. 6 nd Chapter 4 in this volume). Time to event methods also provide different ways of thinking about data, as illustrated by Chapters 2 and 3 in this volume. Data from either a traditional design or a time to event design can be used to estimate LC50, so the goal of the analysis can be the same. The time to event method makes more assumptions (discussed below) and is not as simple as the traditional design. It also requires more data

Table 5.1 Criteria to Compare Statistical Methods

Goals:	What sorts of questions can be answered?
	What quantities are estimated?
Precision:	If estimating the same question, which method is more precise?
Assumptions:	What assumptions are made by the method?
	Is the method robust to violations of those assumptions?
Simplicity:	How difficult is it to do or explain the analysis?
Data requirements:	Do the methods require the same sort of data?

(times of death, not just "dead at time T_o"). However, the benefits of using a time to death model for this purpose are not clear. Just how much more precise is the estimate of LC50 when a time to event model is used?

5.2.1 Estimating LC50 and its precision from a dose response curve

LC50 can be estimated from the slope and intercept of a linear dose-response regression. If the transformed proportion of deaths, logit p_i, at concentration C_i follows a linear dose-response curve,

$$\text{logit } p_i = a_t + b_t\, C_i, \qquad 5.7$$

then the estimated concentration for which $p_i = 0.5$, is

$$\hat{L}C50 = \hat{a}_t/\hat{b}_t,$$

where $\hat{a}_1$ and $\hat{b}_1$ are the estimated intercept and slope. Alternatively, LC50 can be directly estimated using non-linear least squares and the non-linear logit model (Equation 5.6).

The variance of $\hat{L}C50$ and the width of the confidence interval depend on the precision of the dose-response curve, i.e., the variance of $\hat{a}_1$, the variance of $\hat{b}_1$, and their covariance. The variance of $\hat{L}C50$ can be approximated[8] by

$$Var\ \hat{L}C50 \approx \hat{L}C50^2\left(\frac{Var\ \hat{a}_t}{\hat{a}_t^2} + \frac{Var\ \hat{b}_t}{\hat{b}_t^2} - \frac{2\ Cov\ \hat{a}_t,\hat{b}_t}{\hat{a}_t,\hat{b}_t}\right) \qquad 5.8$$

Confidence intervals for LC50 can be derived using Fieller's theorem.[9] Upper and lower bounds are

$$\frac{1}{1-g}\hat{L}C50 + g\frac{Cov\ \hat{a}_t,\hat{b}_t}{Var\ \hat{b}_t} \pm \frac{t_c}{\hat{b}_t}\sqrt{Var\ \hat{a}_t + 2\hat{L}C50\ Cov\ \hat{a}_t,\hat{b}_t + \hat{L}C50^2 \quad Var\ \hat{b}_t - g\left(Var\ \hat{a}_t - \frac{Cov\ \hat{a}_t,\hat{b}_t}{Var\ \hat{b}_t}\right)} \qquad 5.9$$

where $g = (t_c/\hat{b}_t)^2 Var\ \hat{b}_t$ and t_c is the appropriate critical value from a student's distribution.

If a non-linear logit model (Equation 5.6) is used, LC50 is a parameter in the model. The asymptotic variance of LC50 is obtained by inverting the negative matrix of second partial derivatives with respect to the two parameters in the non-linear model, b and LC50. Asymptotic confidence intervals could be constructed as $\hat{L}C50 \pm t_c\sqrt{Var\ \hat{L}C50}$, where t_c is a quantile of a t distribution. These intervals will be symmetric around $\hat{L}C50$, while the

Fieller confidence intervals given by Equation 5.9 will, in general, not be symmetric. In very large samples, the two confidence intervals will be similar, but they may not be in experiments with 50 or 100 individuals.

5.2.2   *Estimating LC50 and its precision from a time to event model*

The fraction of individuals surviving to time t depends on both the dose concentration and t (Equation 5.3). This equation can be rearranged to give the concentration at which the proportion of individuals surviving to time t, $S(t)$, is p:

$$LC_p(t) = \frac{\log t - \hat{\gamma} \; \text{logit} \; p - \hat{\alpha}}{\hat{\beta}}$$

since logit $0.5 = 0$, LC50 can be estimated from the parameters of a log-logistic time to event model as

$$LC50(t) = \frac{\log t - \hat{\alpha}}{\hat{\beta}} \qquad\qquad 5.10$$

Unlike traditional regression methods, which estimate LC50 only for a specific time, time to event models can be used to estimate LC50 for any length of exposure. By comparing coefficients in the time to event model (α and β) to the coefficients in the traditional logit model (a_t and b_t), we see that $\hat{\beta}$ estimates the same quantity as $\hat{b}_t$, but the intercepts ($\hat{\alpha}$ and $\hat{\alpha}_t$) estimate different quantities. The intercept from the time to event model is the intercept from the traditional model shifted by the quantity log t, where t is the time at which survival is recorded (e.g., 96 h). So $\hat{L}C50$ for a specific length of exposure can be estimated by first estimating $\hat{\alpha}$ and $\hat{\beta}$ by maximum likelihood, then submitting those estimates and the length of exposure into Equation 5.10.

The estimation of the variances and covariances of the parameters is a standard part of most censored regression software packages. The variance-covariance matrix of, $\hat{\alpha}$, $\hat{\beta}$ and $\hat{\gamma}$ is approximated by inverting the negative matrix of second derivatives of the log likelihood function.[5] The variance of $\hat{L}C50$ can then be approximated by

$$Var \; \widehat{LC50} \approx LC50^2 \left(\frac{Var \; \hat{a}}{(\hat{a} - \ln t)^2} + \frac{Var \; \hat{b}}{\hat{b}^2} - \frac{2 \; Cov \; \hat{a}, \hat{b}}{\hat{b}(\hat{a} - \ln t)} \right) \qquad 5.11$$

Confidence intervals for LC50 can be estimated using the time to death estimates, variances and covariances in the Fieller method (Equation 5.9). Because the estimated intercept, a, does not directly enter into the equation for the confidence interval, the same confidence interval formula (Equation 5.9) can be used for both regression and time to death models. Although a likelihood can be defined for a non-linear parameterization of the time to death model, the computations cannot be done using widely available software.

5.2.3 Description of example data sets

Statistical models, like the traditional and time to event models, can be compared by choosing some "representative" data sets and computing estimates using each model. The specific value of the estimate (e.g., $\hat{L}C50$) is less important than the precision of the estimate (e.g., variance or confidence interval width). Estimates are random variables, so estimates from different models are unlikely to be the same. For real data, the "right" answer is not known, since the true parameter values are unknown.

Finney[3] used data from five bioassays of poliomyelitis to compare probit regression with a simple time to event method where censored values were replaced by an arbitrary constant. I used one of those data sets to compare the precision of estimates from logit regression and a log-logistic time to event model. Finney reports the number of days between injection with a standard preparation (#21) of poliomyelitis and the appearance of the symptoms. Across all doses, the responses varied from 3 days to > 16 days. The Finney data have relatively crude information on time to death, in the sense that the observed time to death has only three values, 3 days, 4 days or > 4 days. Time to death (3 days or 4 days) was observed on 17 of 50 mice.

Relatively fine scale information on time to death is available in Shepard's[7] data on the effect of low dissolved oxygen on trout survival. Shepard reports observed death times up to 5000 h in waters with different dissolved oxygen concentrations. I increased the number of censored fish by artificially censoring the data at 96 h. Death was observed every h during the initial stages of the experiment. Forty-five of the 100 trout died within the first 96 h.

5.2.4 Exploration of example data sets

The linear dose-response regression (Equation 5.7) was fit to both the Shepard and Finney data using SAS PROC REG.[10] Because the logit transformation is undefined when proportions are either 0 or 1, those values were replaced by $1/2N$ and $1-1/2N$, where N is the number of fish in each dose group. The nonlinear parameterization of the dose-response regression (Equation 5.6) was fit using SAS PROC NLIN.[10] The log-logistic censored time to death model (Equation 5.1) was fit using SAS PROC LIFEREG.[10]

In many toxicological data sets, the time of death is grouped into a relatively small number of groups. For example, in the Finney data, death was recorded daily for 16 days. After censoring, the observations have one of three possible values: 3 days, 4 days, or > 4 days. The typical censored regression model assumes that the time of death is a continuous random variable and can have many possible values. When the number of possible death times is small, it is better to consider the observed death times as interval-censored values.[4] If the Finney data are considered as interval-censored data, there are three possible observations: death between the second and third days, between the third and fourth days, or after the fourth day. A regression model was fit to this interval-censored data using SAS PROC LIFEREG.[10]

LC50, its standard error, and 95% confidence intervals were estimated for all models using the procedures given above. SAS code to fit the models and do the calculations for $\hat{L}C50$ is available from the author.

The $\hat{L}C50$ estimated by each model are not identical, but they are very similar (Table 5.2). Estimates and measures of precision from the linear and non-linear versions of the regression model are very similar. The only difference is a small shift in the endpoints of the confidence intervals. Estimates from time to death models are more precise than estimates from the regression models, when measured either by the standard error of $\hat{L}C50$ or by the width of the confidence interval. The increase in precision can be quantified by the efficiency, defined as:

$$\text{Efficiency} = \left(\frac{s.e.\ \hat{L}C50 \text{ from a regression}}{s.e.\ \hat{L}C50 \text{ from a time to death model}} \right)$$

For these two data sets, the efficiency ranges from 1.49 to 2.96 (Table 5.2).

Table 5.2 $\hat{L}C50$, its Standard Error, and 95% Confidence Intervals Using Linear Regression, Non-Linear Regression and Time-To-Death Models

Data Set	Model	$\hat{L}C50$	s.e.	95% c.i.	Efficiency
Finney	Linear	3.35	0.220	(2.73, 3.81)	
	Non-linear	3.35	0.220	(2.84, 3.86)	
	Time-to-death	3.10	0.153	(2.80, 3.60)	2.06
	Interval TTD	3.26	0.180	(2.92, 3.92)	1.49
Shepard	Linear	1.331	0.043	(1.23, 1.43)	
	Nonlinear	1.331	0.043	(1.23, 1.43)	
	Time-to-death	1.399	0.025	(1.35, 1.47)	2.96
	Interval TTD	1.399	0.025	(1.35, 1.47)	2.96

The linear and nonlinear models are two forms of a dose-response regression using the proportion of survivors at a fixed time. The time-to-death and interval TTD models use the (possibly censored) death times. The Finney and Shepard data sets are described in Section 5.2.3.

Efficiency has a convenient, practical interpretation in terms of the number of animals needed for an experiment. If the efficiency of a particular model is E compared with a reference model (here, linear regression), then the precision using the model with N animals is the same as the precision using the reference model with $N \times E$ animals. Consider the Shepard experiment, which used 100 fish. The precision obtained from using a time to event analysis with those 100 fish is the same as would have been obtained from a traditional analysis of an experiment with 300 fish. Using more animals in an experiment increases the precision of the regression parameters and $\hat{L}C50$, but it (usually) increases the cost of the experiment. When a time to event analysis is more efficient, it provides a way to get the same precision with fewer animals.

5.3 Theoretical precision of time to event and traditional models

Although the time to event model provides more precise estimates for both Finney and Shepard data sets, there is no consistency in the size of improvement. Better understanding of why the improvement is larger in the Shepard data set might come from calculating efficiencies for many data sets and looking for patterns. However, this is difficult because each data set has unique and idiosyncratic properties. A theoretical analysis of the precision provides some general guidelines for when the improvement should be large.

We need a way to compare precision that does not depend on the specific observations in a data set. This can be done using Fisher Information.[5] This is a matrix of expected values of the second partial derivatives of the likelihood with respect to each pair of parameters. Remember, the variances and covariances of the parameter estimates were approximated by inverting the observed matrix of second derivatives. The Fisher Information is the expected value of that matrix, so it does not depend on the specific observations in a data set. It does depend on the model, the true parameter values, and the experimental design (e.g., the set of concentrations used in the experiment).

Fisher Information can be calculated for an entire data set or individually for each observation. Because we will eventually compare the contribution to Fisher Information from a censored observation to the contribution from an exact time of death, it is easier to focus on the information from a single observation.

Imagine that each individual has an unknown "true" time to death, τ_i. In a traditional fixed-time design, we observe one of two possible responses for each individual. At the fixed observation time, T_o, it is dead (i.e., $\tau_i \leq T_o$) or alive (i.e. $\tau_i > T_o$). In a time to death design, we observe either T_i, the time to death, if that is before the end of the experiment (i.e. $\tau_i < T_c$) or "alive at the end of experiment" (i.e. $\tau_i > T_c$). If both experiments last the same length of time (i.e. $T_o = T_c$), live individuals from a traditional experiment are equivalent to live individuals at the end of the time to death experiment. The only difference between the experiments comes from those individuals that die before the fixed recording time. We can quantify the extra precision provided by those individuals using Fisher Information.

Define I^0 as the Fisher Information matrix for one observation where the exact time of death is recorded. Define I^c as the Fisher Information matrix for one right-censored observation (i.e. dead at T_o). These matrices have six unique elements, but we will focus on the three elements related to the dose response parameters, α and β. Analytical solutions for the information matrix entries related to γ, the log-logistic scale parameter, can be written down, but they and their ratios are complicated and not enlightening. Details on the computation of Fisher Information are provided in this chapter's Appendix.

The relative amount of information on an exact time of death (I^0) and "died before T_o" (I^c) can be expressed as a function of the fraction of individuals dying before the end of the experiment, $F(T_0)$, or as a function of the length of the experiment, T_o, relative to the median time to death, T_m.

$$\text{for Var } \hat{\alpha} \qquad \frac{I^0_{\alpha\alpha}}{I^c_{\alpha\alpha}} = 1 + \frac{F(T_0)}{3(1 - F(T_0))} = 1 + \frac{1}{3}\left(\frac{T_0}{T_m}\right)^\gamma$$

$$\text{for Cov } \hat{\alpha}, \hat{\beta} \qquad \frac{I^0_{\alpha\beta}}{I^c_{\alpha\beta}} = 1 + \frac{F(T_0)}{3(1 - F(T_0))} = 1 + \frac{1}{3}\left(\frac{T_0}{T_m}\right)^\gamma$$

$$\text{for Var } \hat{\beta} \qquad \frac{I^0_{\beta\beta}}{I^c_{\beta\beta}} = 1 + \frac{F(T_0)}{3(1 - F(T_0))} = 1 + \frac{1}{3}\left(\frac{T_0}{T_m}\right)^\gamma$$

Things to notice about these ratios are that:

- The ratios are the same for all three elements of the Fisher Information matrix for the dose-response parameters.
- The ratios depend only on $F(T_o)$, the probability of dying before time T_o, the end of the experiment. Alternatively, the information ratios depend on the ratio between T_o and the median time of death, T_m, and γ, the parameter for the spread of the log-logistic distribution.
- $F(T_o)$ is a cumulative distribution function, so it always takes values between 0 and 1. Hence the ratio of information is no smaller than 1. At worst, the information in an observed death time is the same as that in a censored time. The same conclusion can be reached from the alternative parameterization, because the three quantities T_o, T_m, and γ are positive.
- The ratio can be very large if $F(T_o)$ is close to 1. If most individuals exposed to a certain concentration die before the end of the experiment, recording death times provides considerably more information.

- Alternatively, the information ratio is large when the median time to death, T_m, is much shorter than the end of the experiment, T_o, especially if the spread in death times is small (large γ).
- When individuals are exposed to different toxicant concentrations, the median time to death, T_m, and the fraction of surviving individuals, $F(T_o)$, depend on the concentration (unless β, the slope of the dose-response curve, is 0). The total information in the sample is the sum of information provided by each individual. The ratios of total information, for observed times of death and censored times to death, will be a weighted average of the ratios for each concentration.

Information ratios were calculated from each concentration in Shepard's[7] experiment on O_2 tolerance in trout. At the lowest O_2 concentration of 0.77 mg/l, the trout die quickly. The predicted median time to death is 5.2 h and almost all individuals are predicted to die before 96 h (Table 5.4). Consequently, recording the individual as dead at 96 h provides considerably less information than does recording the exact time to death. As the O_2 concentration increases, the predicted median time to death increases and the predicted fraction of fish that die before 96 h decreases. At the highest O_2 concentrations, the predicted median time to death is much longer than 96 h; recording the individual as dead at 96 h provides exactly the same amount of information as does the exact time to death.

The variances of $\hat{L}C50$ and dose-response parameters depend on the total information provided by all observations in the experiment. In a time-to-death experiment, this includes the information from observed times of

Table 5.4 Predicted Median Time to Time to Death, Predicted Fraction Dead at 96 h and Information Ratio for each O_2 Concentration

O_2conc.	Median time to death	Fraction dead at 96 h	Information ratio
0.77	5.2	0.999996	8040
0.94	11.4	0.9994	5342
1.10	23.9	0.99	42
1.16	31.5	0.979	17
1.36	79.1	0.66	1.65
1.43	109	0.39	1.21
1.55	190	0.086	1.03
1.69	362	0.010	1.00
1.77	523	0.003	1.00
1.86	791	· 0.0007	1.00

Information ratio is the ratio of Fisher information for an observed time to death to Fisher information for a censored time. For this analysis, the data were artificially censored at 96 h (= T_o). Parameter estimates were $\hat{\alpha} = -6.56$, $\beta = 16.0$, and $\hat{\gamma} = 3.47$.

and β depends on the correlations between γ and α and β. If γ is estimated independently of α and β, there is no decrease in precision.

The decrease in precision due to the uncertainty in γ can be calculated for specific data sets (Table 5.5). In the Finney data, the correlations between γ and the other two parameters are very close to zero and there is little change in the standard error of the dose-response parameters. In the Shepard data, the correlations between γ and the other two parameters are larger. Incorporating the uncertainty in γ increases the standard error of the dose-response parameters by approximately 3%. If the increase in information is small and the uncertainty in estimating γ is large, estimates from a time to event analysis may be less precise than from a traditional analysis. However, in all the cases examined here, the extra information provided by observed death times more than compensates for the uncertainty in estimating an additional parameter.

5.5 When a time to event model might be less costly than a traditional experiment

The design that provides the least costly way to estimate $\hat{L}C50$ with a desired precision depends on a trade-off between the additional cost to observe time to death and the cost for using additional animals. Observing exact death times may be very costly, especially if it requires continuous overnight observation of a short-term acute study. The cost might be reduced by judicious use of interval-censored observations. For example, in a short-term study, the number of deaths might be observed at 4-h intervals during the day, followed by a 16-h interval overnight. Such interval-censored data provide less information than exact data, but it is still more information than that from a traditional fixed-time design. However, a time to death study may save money because, with the increased precision from a time to death analysis, such a study requires fewer animals. This factor will be more important when animals are more expensive.

Table 5.5 Comparison of Standard Errors of Dose-Response Parameters when Including or Ignoring the Uncertainty in Estimation of γ, the Log-Logistic Scale Parameter

Data Set	Parameter	Correlation with γ	Standard error ignoring γ	incl. γ	Ratio
Finney	α	−0.02	0.109	0.109	1.00
	β	0.06	0.0267	0.0268	1.00
Shepard	α	−0.25	0.325	0.336	1.03
	β	0.27	0.294	0.306	1.04

5.6   Conclusions

This chapter has compared time to event analyses with traditional probit/logit analyses. The traditional probit/logit analysis can be viewed as an analysis of doubly censored time to event data. This means that both models are estimating the same fundamental parameters. Hence, it makes sense to compare the relative precision of dose-response parameters and $\hat{L}C50$ from each analysis. Numerical examples show that the time to death models usually, but not always, provide more precise estimates. A theoretical analysis of Fisher Information for a log-logistic model with linear dose response shows that an observed death time provides more information than a traditional observation of "dead at the end of the experiment." The amount of the improvement depends on either the probability that an individual dies by the end of the experiment or the ratio of the length of the experiment and the median time to death. The increase in information translates into more precise estimates of LC50 and dose-response parameters, if the uncertainty in estimating the log-logistic shape parameter is ignored. This uncertainty or violation of model assumptions may lead to estimates from time to event models that are less precise than those from traditional analyses. However, traditional fixed time designs provide information about one specific length of exposure. Time to event data and analyses provide more complete information about response to toxicants. They can be used to describe how mortality varies with both dose and length of exposure.

Acknowledgments

Research and manuscript preparation were partially supported by Financial Assistance Award DE-FC09-96SR18546 from the U.S. Department of Energy to the University of Georgia Research Foundation. A grant from SETAC (UK) to support the author's participation in the April 1998 meeting that led to this book was very much appreciated, as were the insightful comments from participants at that meeting.

Appendix

Here the details of the likelihood function and Fisher Information for a single observed time to death and for a single left-censored time to death (i.e., an observation from a traditional fixed time experiment) are presented. The statistical model is that times to death follow a log-logistic distribution with a mean log time to death that is a linear function of concentration (e.g., $\mu_i = \alpha + \beta C_i$). The spread of log-transformed times to death is the same at all concentrations. I will also assume that a traditional experiment is performed for the same length of time as the time to death experiment. This means that the censoring time in the time to death experiment is the same as the fixed time at which alive or dead is recorded in the traditional experiment.

The following notation will be used:

α intercept of the dose-response curve
β slope of the dose-response curve
C_i concentration to which individual i was exposed
$f(t)$ probability of dying at time t
$F(t)$ probability of dying at or before time t
μ_i median time to death for individual i
$l_c(t)$ log likelihood for a single censored (at time t) time to death
   $l_o(t)$log likelihood for a single observed time to death at time t
τ_i time to death for individual I
T_i observed time to death in a time to death experiment
T_c censoring time in a time to death experiment
T_m median time to death in a time to death experiment
T_o observation time in a traditional experiment

The contribution to the log likelihood for a single observed value, T_i, is

$$l_o(T_i) = \log \gamma - \gamma \log T_i + \alpha + \beta C_i - \log T_i - 2 \log (1 + e^{-\gamma \log T_i + \alpha + \beta C_i}),$$

while that for a left-censored time to death $(\tau_i \leq T_o)$ is

$$l_c(T_i) = -\log(1 + e^{-\gamma \log T_i + \alpha + \beta C_i}).$$

The Fisher information matrix is the negative expected value of the matrix of second derivatives.[5] Because we are interested only in comparing information for individuals that die before time T_o, each element of the information matrix is the solution of an integral like

$$I_{aa} = \frac{-1}{F(T_0)} \int_{t=0}^{T_0} \frac{\partial^2 l}{\partial \alpha^2} f(t) dt.$$

The symbolic math package Maple[11] was used to evaluate the second derivatives and integrals. For α and β, the two parameters of the dose-response curve, the second derivatives and elements of the information matrix for observed times to death, t, are:

$$\frac{\partial^2 l_0(t)}{\partial \alpha^2} = -2 \frac{e^{\alpha + \beta C_i}(e^{2(\alpha + \beta C_i)} t^\gamma + 2e^{\alpha + \beta C_i} t^{2\gamma} t^3 \gamma)}{(t^\gamma + e^{\alpha + \beta C_i})^4}$$

$$\frac{\partial^2 l_0(t)}{\partial \alpha \partial \beta} = -2C_i \frac{e^{\alpha + \beta C_i}(e^{2(\alpha + \beta C_i)} t^\gamma + 2e^{\alpha + \beta C_i} t^{2\gamma} t^3 \gamma)}{(t^\gamma + e^{\alpha + \beta C_i})^4}$$

$$\frac{\partial^2 l_0(t)}{\partial \beta^2} = -2C_i^2 \frac{e^{\alpha + \beta C_i}(e^{2(\alpha + \beta C_i)} t^\gamma + 2e^{\alpha + \beta C_i} t^{2\gamma} t^3 \gamma)}{(t^\gamma + e^{\alpha + \beta C_i})^4}$$

$$I_{\alpha\alpha}^0 = \frac{T_0^{3\gamma} + 3T_0^{2\gamma}e^{\alpha+\beta C_i}}{3F(T_0)(T_0^\gamma + e^{\alpha+\beta C_i})^3}$$

$$I_{\alpha\beta}^0 = \frac{C_i(T_0^{3\gamma} + 3T_0^{2\gamma}e^{\alpha+\beta C_i})}{3F(T_0)(T_0^\gamma + e^{\alpha+\beta C_i})^3}$$

$$I_{\beta\beta}^0 = \frac{C_i^2(T_0^{3\gamma} + 3T_0^{2\gamma}e^{\alpha+\beta C_i})}{3F(T_0)(T_0^\gamma + e^{\alpha+\beta C_i})^3}$$

The corresponding quantities for left-censored observations (i.e., dead at T_o) from a traditional design are:

$$\frac{\partial^2 l_c(t)}{\partial \alpha^2} = -\frac{t^\gamma e^{\alpha+\beta C_i}}{(t + e^{\alpha+\beta C_i})^2}$$

$$\frac{\partial^2 l_c(t)}{\partial \alpha \partial \beta} = -\frac{C_i t^\gamma e^{\alpha+\beta C_i}}{(t + e^{\alpha+\beta C_i})^2}$$

$$\frac{\partial^2 l_c(t)}{\partial \beta^2} = -\frac{C_i^2 t^\gamma e^{\alpha+\beta C_i}}{(t + e^{\alpha+\beta C_i})^2}$$

$$I_{\alpha\alpha}^c = \frac{T_0^{2\gamma} + e^{\alpha+\beta C_i}}{F(T_0)(T_0^\gamma + e^{\alpha+\beta C_i})^3}$$

$$I_{\alpha\beta}^c = \frac{C_i T_0^{2\gamma} + e^{\alpha+\beta C_i}}{F(T_0)(T_0^\gamma + e^{\alpha+\beta C_i})^3}$$

$$I_{\beta\beta}^c = \frac{C_i^2 T_0^{2\gamma} + e^{\alpha+\beta C_i}}{F(T_0)(T_0^\gamma + e^{\alpha+\beta C_i})^3}$$

The second derivatives involving γ, the scale parameter for the log-logistic distribution are complicated and expressions for the information in an exact observation and a censored observation do not simplify nicely.

The ratios of the information for dose-response parameters do simplify nicely:

$$\frac{I_{\alpha\alpha}^0}{I_{\alpha\alpha}^c} = 1 + \frac{T_0^\gamma}{3e^{\alpha+\beta C_i}}$$

$$\frac{I_{\alpha\beta}^0}{I_{\alpha\beta}^c} = 1 + \frac{T_0^\gamma}{3e^{\alpha+\beta C_i}}$$

$$\frac{I^0_{\beta\beta}}{I^c_{\beta\beta}} = 1 + \frac{T^\gamma_0}{3e^{\alpha+\beta C_i}}$$

These can be further simplified using properties of the log-logistic distribution (Equations 5.2–5.4). The quantity $T^\gamma_0/e^{\alpha+\beta C_i}$ is equal to $F(T_0)/(1 - F(T_0))$, leading to expressions for the ratio of information in terms of the fraction of individuals dying during the experiment.

$$\frac{I^0_{\alpha\alpha}}{I^c_{\alpha\alpha}} = 1 + \frac{F(T_0)}{3(1 - F(T_0))}$$

$$\frac{I^0_{\alpha\beta}}{I^c_{\alpha\beta}} = 1 + \frac{F(T_0)}{3(1 - F(T_0))}$$

$$\frac{I^0_{\beta\beta}}{I^c_{\beta\beta}} = 1 + \frac{F(T_0)}{3(1 - F(T_0))}$$

Alternatively, the median time to death at any concentration, T_m, obtained by solving $F(T_m) = 0.5$, is a function of the parameters and the exposure concentration.

$$T_m = (e^{\alpha+\beta C_i})^{\frac{1}{\gamma}}$$

Substituting T^γ_m for $e^{\alpha+\beta C_i}$ in the information ratios gives a simplification in terms of the median time to death:

$$\frac{I^0_{\alpha\alpha}}{I^c_{\alpha\alpha}} = 1 + \frac{1}{3}\left(\frac{T_0}{T_m}\right)^\gamma$$

$$\frac{I^0_{\alpha\beta}}{I^c_{\alpha\beta}} = 1 + \frac{1}{3}\left(\frac{T_0}{T_m}\right)^\gamma$$

$$\frac{I^0_{\beta\beta}}{I^c_{\beta\beta}} = 1 + \frac{1}{3}\left(\frac{T_0}{T_m}\right)^\gamma$$

References

1. Newman, M.C., *Quantitative Methods in Aquatic Ecotoxicology*, Lewis, Chelsea, MI, 1995.
2. Hamilton, M.A., Estimation of the typical lethal dose in acute toxicity studies, in *Statistics in Toxicology*, Krewski, D. and Franklin, C., Eds., Gordon and Breach Science Publishers, New York, NY, 1991, 61.

3. Finney, D.J., *Statistical Method in Biological Assay*, Charles Griffin and Co., London, UK, 1978.

4. Dixon, P.M. and Newman, M.C., Analyzing toxicity data using statistical models of time to death: an introduction, in *Metal Ecotoxicology: Concepts and Applications*, Newman, M.C. and McIntosh, A.W, Eds., Lewis, Chelsea, MI, 1991, 207.

5. Edwards, A.W.F., *Likelihood*, Cambridge University Press, Cambridge, UK, 1972.

6. Sun, K., Krause, G.F., Mayer, F.L., Jr., Ellersieck, M.R. and Basu, A.P., Predicting chronic lethality of chemicals to fishes from acute toxicity test data: theory of accelerated life testing. *Environ. Toxicol. & Chem.*, 14, 1745, 1995.

7. Shepard, M.P., Resistance and tolerance of young speckled trout (*Salvelinus fontinalis*) to oxygen lack, with special reference to low oxygen acclimation. *J. Fish. Res. Bd. Can.*, 12, 387, 1995.

8. Mood, A.M., Graybill, F.A. and Boes, D.C., *Introduction to the Theory of Statistics*, 3rd ed., McGraw-Hill, New York, NY, 1974.

9. Piegorsch, W.W. and Bailer, A.J., *Statistics for Environmental Biology and Toxicology*, Chapman and Hall, London, UK, 1997.

10. SAS Institute, *SAS/STAT User's Guide, version 6*, 4th ed., Cary, NC, 1990.

11. Heck, A., *Introduction to Maple*, 2nd ed., Springer-Verlag, New York, NY, 1993.

chapter 6

Using time to event modeling to assess the ecological risk of produced water discharges

Chris C. Karman

Contents

6.1 Introduction

Since 1991, the North Sea countries (the United Kingdom, the Netherlands, Norway and Denmark) have been developing a decision support system to legislate the use and discharge of offshore exploration, drilling and production chemicals. The heart of this "Harmonized Mandatory Control System" is the CHARM model (Chemical Hazard Assessment and Risk

Management). This model enables the ranking of chemicals on the basis of their intrinsic properties, using a realistic worst-case scenario (i.e., 95 percentile of known field conditions occurring in the North Sea). To meet the prerequisites of the model (simple and transparent calculation rules), the CHARM model uses a fixed dilution factor, assuming equal and constant dispersion of chemicals around the platform. Although such a model is suitable for an internationally harmonized control system, a more realistic prediction is required to assess the actual risk of produced water. Because of increasing environmental concerns about the impacts of oil and gas exploration and production, and a shift toward self-regulation, the Norwegian Exploration and Production Industry initiated the development of a more detailed risk assessment model. In this model (DREAM: Dose Related Effect Assessment Model), environmental risk is dependent on the type of organism and its sensitivity as a function of exposure time. The DREAM project began in 1998 and is expected to be completed in 2001. Most effort will be put into the development of time to event models for representative chemicals and organisms. These models will be derived using the results of an extensive experimental program.

Since the actual development of DREAM was predicted to take approximately 4 years, the Dutch Oil Company (NAM B.V.) asked the Netherlands Organization for Applied Scientific Research (TNO) to develop a rapid and simpler risk assessment model to include time to event methodology. This model (DRACO: Dynamic Assessment of Chemicals Discharged Offshore) gives a probabilistic estimation of the ecological risk of produced water in which spatial and temporal variation in the concentration of chemicals is summarized in frequency distributions, and aims to support the selection of cost-effective mitigating measures for risk reduction.

6.2 *Exposure models for offshore oil and gas exploration and production*

In the offshore production of oil and gas, production chemicals are added to the water injected into the well to enhance production, to protect the equipment and to maintain safety on the platform. A proportion of these chemicals may end up in the marine environment as they are discharged with the produced water, and adverse effects to the biota in the ambient environment may result. To prevent pollution caused by exploration and production activities, each North Sea country has developed a legislative system for the use of chemicals in these activities. As the systems have been developed individually, regulation has become inconsistent between the North Sea countries. The North Sea Ministers Conference in 1990 therefore decided to harmonize legislation on the use of offshore exploration and production chemicals. In 1991, the CHARM project was initiated, in which authorities of the North Sea countries, the chemical suppliers and operating companies cooperated in the development of a model for evaluating the environmental impact of the discharge of offshore Exploration and Production (E&P) chemicals. The CHARM model[1] estimates the concentration of a chemical in the water 500m from the platform, based upon worst-case platform characteristics (i.e., 95 percentile of platform characteristics on the basis of all platforms in the North Sea region). This concentration is then compared with an ecosystem toxicity threshold, extrapolated from acute toxicity data. The CHARM model is accepted by the Paris Commission (PARCOM) as a valid model for performing hazard assessment of offshore E&P chemicals, as required by the Harmonized Mandatory Control System (the regulatory framework for each of the North Sea countries).

To estimate the concentration of a chemical at 500 m from a platform, the concentration of a chemical in the discharged produced water is multiplied by a fixed dilution factor. In the Hazard Assessment module of the CHARM model (using realistic worst-case assumptions) a default dilution of 1:1000 is used, although experimentally determined figures may be used as well.[2,3] The spatial and temporal distribution of the chemical is not taken into account in either case, and a constant exposure concentration of the chemical is assumed at a fixed distance from the platform. In reality, however, the produced water will form a plume within the ambient seawater, with a high concentration at the center of the plume and declining concentrations as the distance from the center and the distance from the discharge point increase (Figure 6.1). Furthermore, the position of the plume may change over time due to influence of tidal current movement and wind. Figure 6.2 shows the concentration of a chemical in the produced water over time during a 24-h discharge of produced water in a tidal area. Because of the spatial and temporal variations in the environmental concentration, and with differences in the behavioral pattern of different species, the exposure is far from constant.

The offshore E&P industry has recognized the need for a more accurate assessment of the environmental risk of produced water discharges. Growing environmental concerns have led to a shift from governmental regulation to self-regulation of the E&P industry, in which industry is expected to take responsibility for increasing environmental care to levels exceeding those required by legislation. These changes have given rise to a need for more accurate risk assessment models.

In particular, traditional risk assessment models are based on the assumption of exposure to a fixed concentration during a certain period of time. An explicit acknowledgment of temporal and spatial variation would lead to more realistic risk assessment models. This chapter describes two such approaches. The first is DREAM, a model under development since 1998 and nearing completion. The second approach is DRACO, which was developed as a pragmatic alternative to DREAM, and which is described in detail here.

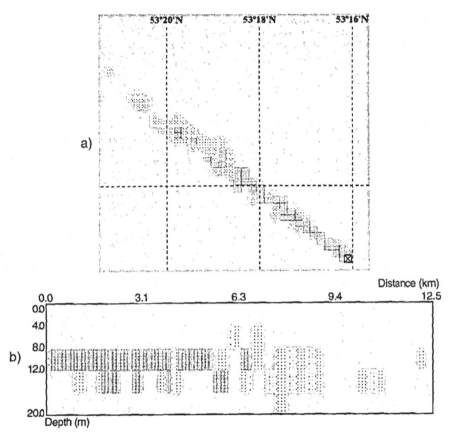

Figure 6.1 Concentration pattern of a plume of produced water continuously discharged from an offshore gas production platform: a) maximum concentration seen from the top; b) concentrations in a cross section of the water layer in the direction of the plume. The figures are generated using a software package suitable for 3-dimensional modeling of discharges into the marine environment.[4]

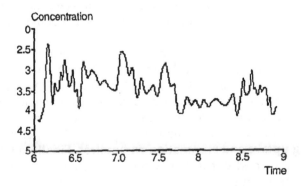

Figure 6.2 Plot of the concentration of a chemical at a random point in the discharge plume shown in Figure 6.1 over a 24-h period.

6.2.1   Dose related effect assessment model (DREAM)

The objective of the DREAM model is to estimate the environmental risk of produced water discharges, emphasizing long-term exposure to low concentrations of chemicals, time-variable concentrations and mixtures of chemicals. The model will be implemented in a user-friendly computer application based on a graphical user's interface. The basis for the model is PROVANN,[4] a computer application for running scenarios with a three-dimensional dispersion model. A program designed for the development of DREAM will take place (Figure 6.3), involving a joint research project among the Norwegian oil companies (Statoil, Norsk Hydro, Elf, and Agip) and several research institutes (TNO in the Netherlands, and Sintef, Rogaland Research and Allforsk in Norway).

The PROVANN model has frequently been applied in case studies for the offshore E&P industry and field validation studies have proven that it is capable of generating a realistic estimation of the environmental concentrations of chemicals discharged with produced water.[4] In the DREAM project, this model is being extended with a module that estimates both the external and internal exposure concentration of organisms such as zooplankton, phytoplankton, benthic organisms and fish. These organisms have been chosen to represent groups of organisms passively moving within the water body (exposed to a decreasing environmental concentration — Figure 6.4a), sessile organisms (exposed to a discharge plume moving with the tide — Figure 6.4b), and actively moving organisms (exposed to high concentrations when swimming into the plume and low concentrations when swimming out of the plume — Figure 6.4c). On the basis of exposure concentrations and dose-response curves for the specific substances, the environmental impact (defined as the effect on a selected species) of produced water discharges can be estimated. Because the exposure concentration may be very variable, the dose-response curve should include the factor of time. For this reason, for the most relevant combinations of species and substances, time to event models are being developed that can be regarded as a dose-response curve, having time as the third dimension. The development of relevant and useful time to event models for risk prediction is one of the most challenging parts of the DREAM project.

Until now, most toxicity testing of produced water compounds have been carried out using standard test protocols with fixed exposure times (48-h *Acartia* test, 72-h algae test, and 96-h fish test). During the DREAM project, an extensive experimental program is being carried out to deliver the necessary data for developing time to event models. Apart from effect studies carried out at a broad range of exposure times, effort is being put into deriving uptake and elimination parameters. Finally, a field validation study is being carried out to verify model calculations with results from semi-field experiments.

The DREAM model is expected to generate a "real-time" environmental impact value (i.e., impact may increase or decrease over time), most probably

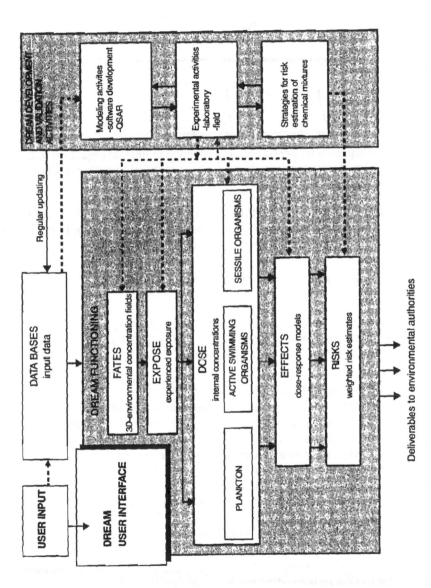

Figure 6.3 Overview of the DREAM-concept (left) and the corresponding main activities (right) required for development of the computer model for risk assessment of time-varying exposure to chemical mixtures in aquatic environments.

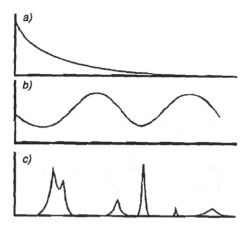

Figure 6.4 External exposure concentration, as a function of time, of organisms with different behavioral patterns: a) passively moving organisms; b) sessile organisms; c) actively moving organisms.

based upon the effects expected in the model species. Extrapolation using QSAR approaches in addition might widen the scope of the model to other species of particular relevance for the region under study.

6.2.2 *Dynamic risk assessment of chemicals discharged offshore (DRACO)*

DRACO is a model capable of dealing with time-variable exposure to chemicals discharged with produced water. DRACO follows a pragmatic, empirical approach, which will be described in the following paragraphs.

6.2.2.1 *Time-integrated exposure concentration*

Figure 6.2 shows that the concentration of a chemical at a certain point within a plume of produced water can be very variable and unpredictable. The variation in concentration is caused by factors such as wind and currents, which might, in turn, cause complicated turbulence patterns in the water column. A sessile organism located at this point also experiences these variations in the concentration of contaminants present. It is clear that an accurate hazard assessment should acknowledge these variations instead of assuming a constant exposure concentration. The relationship between concentration and time is too unpredictable to be described by a mathematical function. It is, however, possible to split up the exposure period into very short periods in which the variation in exposure concentration can be considered negligible. If the concentration range is divided into a series of (logarithmic) classes, a time distribution pattern can be generated by accumulating the actual time that the exposure concentration experienced at a fixed point lies within a specific concentration class. Assuming that the effect of repetitive short exposure (i.e., 4 x 6 hours) equals the effect of continuous exposure for a period

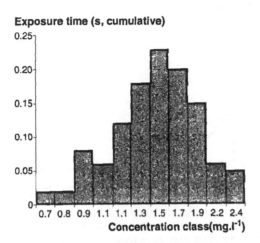

Figure 6.5 Time distribution of the concentration curve presented in Figure 6.2. The
x-axis represents the concentration range divided into logarithmic classes. The y-axis
represents the cumulative exposure time to the specified concentration class.

as long as the short exposure periods together (i.e., $4 \times 6 = 24$ hours), the
concentration-time curve can be replaced by the time-distribution curve.
Figure 6.5 shows the distribution of cumulative exposure time per concentration
class, generated from the concentration-time curve presented in Figure 6.2.

To perform a hazard assessment that conforms with the PEC:PNEC
approach (PEC: predicted environmental concentration; PNEC: predicted no
effect concentration), each concentration class can be compared with a predicted no effect concentration, corresponding to the cumulative exposure
time of that concentration class.

6.2.2.2 *Time-adjusted effect concentration*

For most chemicals used on offshore production platforms ecotoxicity data
have to be provided for regulatory purposes.[5] These are usually time-specific
LC50 or EC50 values for algae, crustaceans or fish determined in standardized toxicity tests. In these tests, organisms are exposed to a constant concentration of the chemical for a predetermined period (e.g., 48 h for the
crustacean *Acartia tonsa*, 72 h for the alga *Skeletonema costatum* and 96 h for
most fish larvae tests) but, as shown in Figure 6.5, organisms in the field are
seldom exposed to a constant concentration for that long. To enable a valid
comparison of the exposure concentration with a no effect concentration, the
latter should be determined for a representative exposure period. Because
only standardized ecotoxicity data (reported as time-specific summaries of
toxicity) are usually available for a chemical, these data need to be adjusted
for a shorter (or longer) exposure time.

The theoretical relationship between exposure time and effect concentration
was first described in 1924 by Haber,[6] who studied the effect of different times

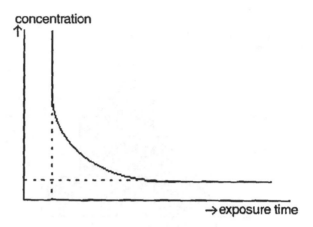

Figure 6.6 Relationship between the concentration at which a defined effect is observed and the corresponding exposure time.

of exposure to poisonous gases on cats. The relationship assumes that a very high concentration is needed over a short exposure time to achieve the same effect as long exposure to a low concentration (see Figure 6.6). Although this theory may not be applicable to all chemicals (e.g., according to Terracini,[7] carcinogens behave differently) a general relationship can be determined for most chemicals, either mechanistically[8,9] or deterministically.[10,11] In this chapter, a deterministic approach, which is relatively simple but describes the relation between exposure-time and effect, will be used:[12]

$$LC50_{/t} = LC_{50,T}/(t/T) \qquad\qquad 6.1$$

where t = actual exposure time; $LC50_{/t}$ = lethal concentration for 50% of the exposed individuals at the actual exposure time t; T = exposure time used in the toxicity test; $LC50_T$ = lethal concentration for 50% of the exposed individuals as determined in the toxicity test using exposure time T.

Equation 6.1 can also be used to calculate a time-adjusted EC50 or PNEC by replacing LC50 in the equation by these figures.

6.2.2.3   Risk calculation

The previous sections show how to deal with fluctuating exposure concentrations by calculating the concentration-time distribution. An organism located at the point shown in Figure 6.2 is exposed to a concentration within each concentration class for a total period as shown in Figure 6.5. The exposure time per class could be used to calculate a representative PNEC value for each class in the distribution, as was shown in the previous section. The ecological risk of being exposed to fluctuating concentrations can now be calculated using the following steps:

1. Each class in the distribution can be represented by the geometric mean of the upper and lower boundaries (referred to as PEC_{class}). For

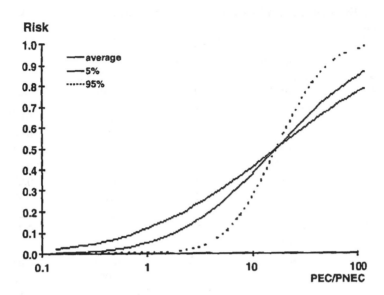

Figure 6.7 The relationship between PEC:PNEC ratio and an ecological risk estimate, calibrated using toxicity data of 17 chemicals.

each class, a PEC:PNEC ratio can be calculated by dividing PEC_{class} by the PNEC adjusted for the cumulative exposure time of that class.

2. To obtain a PEC:PNEC ratio for the overall period of exposure, the ratios of the individual classes need to be combined. Since the addition of the ratio will lead to an overestimation of the environmental risk, the PEC:PNEC ratio per class is transformed into a probabilistic risk estimate, using the relationship presented in Figure 6.7.

3. Finally, the risk estimates for each class can be combined into a single overall risk estimate for the period of exposure to the fluctuating concentration shown in Figure 6.5. As the risk estimate is a probability value and the risk estimates for the individual concentration classes are independent, the statistical rule for combining probabilities (Equation 6.2) can be used to combine the risk estimates.

$$R(c1+c2) = R(c1)+R(c2) - R(c1) \, R(c2) \qquad\qquad 6.2$$

where R(c1+c2) = environmental risk of a mixture of 1 and 2, R(c1) = environmental risk of compound (or mixture) 1, R(c2) = environmental risk of compound (or mixture).[2]

These three steps can be repeated for each cell in the distribution pattern; Table 6.1 shows an example of the calculation steps. Finally, this yields a (time-integrated) risk estimate for a three-dimensional pattern of cells, which can be displayed as shown in Figure 6.8.

Table 6.1 Example Calculations Performed with the Model for the Time Variable Exposure as Shown in Figure 6.2

Class (mg l⁻¹)	Cumulative exposure time(s)	Time adjusted PNEC (mg l⁻¹)	PEC:PNEC ratio	Risk
0.7	0.02	128	0.005	0.00%
0.8	0.02	128	0.006	0.00%
0.9	0.08	26	0.035	0.01%
1.0	0.06	32	0.032	0.00%
1.1	0.12	16	0.072	0.03%
1.3	0.18	11	0.122	0.11%
1.5	0.23	9	0.173	0.24%
1.7	0.20	10	0.170	0.23%
1.9	0.15	13	0.148	0.18%
2.2	0.06	32	0.067	0.03%
2.4	0.05	43	0.057	0.02%
			Combined risk:	0.85%

6.2.2.4   *Sensitivity analysis*

It is apparent that the model result is dependent on the number of classes used to calculate the time-integrated risk. A sensitivity analysis was carried out to demonstrate the influence of the number of classes (defined for the concentration pattern presented in Figure 6.2) on the final result of the DRACO model. The result of this analysis is presented in Figure 6.9, which shows that the model is most sensitive for the number of classes in the range from 1 to 10 concentration classes. Furthermore, in this figure, the results of the sensitivity analysis are compared with the result of a traditional PEC:PNEC risk calculation. This leads to the observation that the time-integrated approach of DRACO enables a more refined risk assessment, preventing a conservative estimate such as that generated using a worst-case PEC concentration from the time profile (i.e., the 95 percentile concentration). However, the approach is more conservative than taking only the median (i.e., geometric mean) concentration from the profile and disregarding the peak values.

6.3   Conclusions

After several years of risk assessment models for discharges from oil and gas production platforms being developed, models in which the exposure time is no longer a fixed value but a variable parameter are needed. Time to event models, which can be regarded as three-dimensional time-dose-response curves, seem to be useful in fulfilling this need in environmental risk assessment.

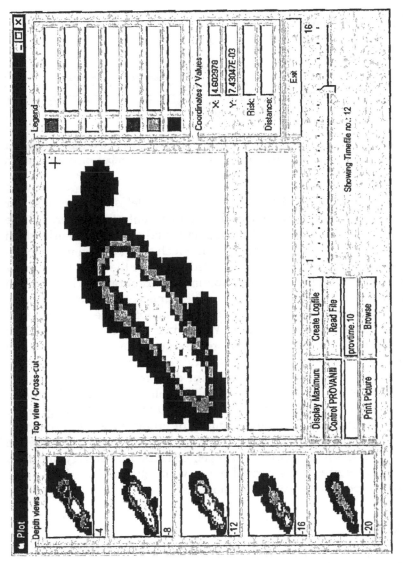

Figure 6.8 Graphical representation of the risk of approximately 20 h exposure to a plume of produced water (see Figure 6.1). The small figures on the left are the average concentration in a depth-layer of the plume. This program distinguishes five depth layers. The large figure in the center is an enlargement of the second depth layer from the surface area.

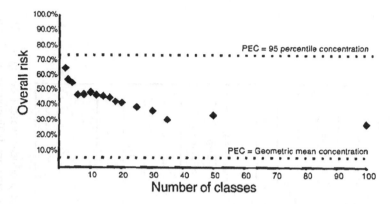

Figure 6.9 Results of the uncertainty analysis carried out with the time-integrated PEC:PNEC approach showing risk as a function of the number of concentration classes.

Although not yet validated, the DRACO model shows promise for implementation in risk assessment of dynamic concentration patterns. Since DRACO is an empirical model, it is pragmatic and straightforward, which makes it applicable in various situations in which the variation in the exposure concentration is measured or modeled, without the need to generate additional data through extensive experimental programs. The DREAM model is an example of the application of mechanistic time to event models and is expected to generate a more realistic estimation of the environmental impact of produced water discharges. When data have been generated for this specific field of application (offshore production of oil and gas) it can be applied in a broad range of situations. Because the number of assumptions is greatly reduced and much more relevant data are utilized, the model's results may be of much more value than the results of the DRACO model.

Although it is not yet clear whether DRACO or DREAM will give the best prediction of the environmental impact of produced water discharges, it has become apparent that time to event models will greatly increase the value of risk assessment studies. However, combined laboratory and field validation studies need to be carried out to verify the precise added value of more advanced time to event risk modeling.

References

1. Karman, C.C., Vik, E.A., Schobben, H.P.M., Ofjord, G.D., and van Dokkum, H.P., *CHARM III. Main Report.* TNO-report R96/355, Den Helder, the Netherlands, 1996.
2. Reerink, H.G. and Verstappen, P., *Quantification of Environmental Effects of Chemicals in Offshore Produced Water: Location Specific Factors,* NAM Report, the Netherlands, 1996.

3. Reerink, H.G. and Verstappen, P., *Sensitivity Analysis on Environmental Effects for NAM Offshore Produced Water Using Location Specific Factors in CHARM*, NAM Report, the Netherlands, 1996.

4. Reed, M., Johnsen, S., Melbye, A. and Rye, H., PROVANN. A model system for assessing potential chronic effects of produced water, in *Produced Water 2. Environmental Issues and Mitigation Technologies*. Environmental Science Research Vol. 52, Reed, M., Johnsen, M., Eds., Plenum Press, New York, NY, 1996, 317.

5. PARCOM., *Decision 96/3 on a Harmonized Mandatory Control System for the Use and Reduction of the Discharge of Offshore Chemicals*. Summary Record of the Joint Meeting of the Oslo and Paris Commissions 1996, 1996.

6. Haber, F., *Fünf Vorträge aus Jahren 1920-1923*. Springer-Verlag, Berlin, 1924, 76.

7. Terracini, B., Cancer hazard identification and qualitative risk assessment. *The Science of the Total Environment*, 184, 91, 1996.

8. Sprague, J.B., Measurement of pollutant toxicity to fish. I. Bioassay methods for acute toxicity, *Water Research*, 3, 793, 1969.

9. French, D.P., Estimation of exposure and resulting mortality of aquatic biota following spills of toxic substances using a numerical model, in *Aquatic Toxicology and Risk Assessment: 14th volume*, Mayes, M.A. and Barron, M.G., Eds., ASTM STP 1124, American Society for Testing and Materials, Philadelphia, PA, 1991, 35.

10. Kooijman, S.A.L.M., Parametric analyses of mortality rates in bioassays, *Water Research*, 15, 107, 1981.

11. Scholten, M.C.Th., Schobben, H.P.M., van het Groenewoud, H., Karman, C.C., Adema, D.M.M. and Dortland, R.I., *An Evaluation of the Regulations for the Control of the Discharge of Noxious Liquid Substances from Ships (Annex II of MARPOL 73/78), Based on an Ecotoxicological Risk Analysis*. TNO Report R93/287, Den Helder, the Netherlands, 1993.

12. Karman, C.C., Schobben, H.P.M. and Scholten, M.C.Th., 1995, *A Description of Recently Developed Methods for Ecological Impact Analysis*, TNO report R95/239, Den Helder, the Netherlands, 1995.

chapter 7

Time to event analysis in the agricultural sciences

John S. Fenlon

Contents

7.1   Introduction

In this chapter, the area of insect pathology is used to demonstrate an approach to time to event analysis that has much in common with ecotoxicology. Insect pathology involves the study of mortality when insects are challenged by pathogenic microorganisms. The time to response of such studies is of particular interest because pathogens (biopesticides) do not have the immediacy of chemical control agents. Two examples are used to demonstrate several methods used in insect pathology for analyzing time to response. The examples illustrate both a simple parametric analysis, and the use of survival analysis in a semi-parametric context. It is shown that, although a modeling approach is used, the methods are robust and have considerable potential for routine time to response analysis in ecotoxicology.

7.2 Temporal events in agriculture

Many processes in agriculture rely on development rates, and are thus intimately associated with time. In attempting to understand these processes, a need to develop experimental procedures that have relied heavily on time to response analyses has arisen. Some specific examples of time to event in horticulture (in particular) are the following:

- The germination of seeds
- Time to flowering or ripening
- Growth rates of mycelium
- Time to emergence (insect pests or diseases)
- Time to death of pests when controlled by pesticides

The first two examples relate directly to primary production. In the case of time to flowering, the response is the culmination of a long process that may or may not be easily understood. The rate of development of a crop is dependent on many processes that interact in complex ways. For example, rate of development will vary from one region to another as a result of climate and soil type, but it may also vary within a region. Indeed, variation may occur within the same field due to changes in soil moisture or soil condition as well as microclimate. From a scientific point of view, the analysis of time to flowering may be purely empirical, particularly in comparative experiments where different cultivars or husbandry techniques (e.g., protection or irrigation systems) are being compared. An important factor to bear in mind, however, is that not only is the measurement of the time to event important, but so is the variability of that time.

Germination of seeds can be considered as a process that is driven by two underlying variables: temperature and water availability, both of which are limiting.[1] Temperature is the major driving factor, and, when water is not limiting, the rate of germination of seeds is strongly driven by temperature. The germination rate is approximately linearly related to temperature up to an optimum temperature, beyond which the rate falls away quite dramatically. Additional complications relate to the fact that any batch of seeds will contain a proportion of non-viable seeds. These should be independent of temperature, but there will also be further seeds that will not germinate on a temperature-dependent basis. If temperature is not limiting, but water status is, the process can be terminated. Mathematically, such systems can be modeled using harmonic response models[2] in which double reciprocal functions ensure that the process stops if one of the sub-processes becomes limiting.

Time to emergence of insect pests has been a significant factor in developing more efficient use of pesticides and in understanding the implications of global warming on pest populations. The life cycle of insects is strongly mediated by temperature, and it has been possible to develop stochastic simulation models,[3] which predict the time of emergence of several groups of economically important agricultural pests.

An alternative to the use of chemical pesticides occurs in the development of biological control agents such as viruses, bacteria and fungi that have specific activity on many agricultural and horticultural pests.[4] These methods differ substantially from conventional pesticides in several respects — first, they do not tend to kill the insect directly, as does the topical application of a chemical pesticide, and second, they usually operate through infectivity rather than toxicity. As with traditional chemical-based assays, there is an interest in the level of pesticide required to kill the target insects, but frequently of more importance is the knowledge of the time required to kill or incapacitate the pest. In the development that follows, the examples will be taken from this field of application. The examples relate to quantal response data,[5] i.e., the response, usually binary, of individuals to a stimulus. There are fewer examples in insect pathology of quantitative responses, although, for sublethal responses of a toxin, one could imagine measurements such as weight gain, fecundity and longevity.

7.3   Historical perspective

The analysis of time-mortality data was originally considered by Bliss[6] and expanded in three papers by Sampford.[7-9] Bliss considered the relatively simple case in which mortality reached 100% and proceeded to a straightforward analysis of the grouped frequencies using a normal or lognormal model. He stressed the importance of testing the assumption of normality and suggested the use of a probit-time plot: if the points were more or less linear, the normality assumption was justified. Litchfield[10] developed this idea and proposed a graphical solution to the estimation of LT50 — time to 50% mortality. Despite warnings by Sampford, this method was fairly widely adopted and Sampford's own analytical solutions widely ignored.

The introduction of Generalized Linear Models by Nelder and Wedderburn,[11] and a reexamination of the statistics of life tables by Cox,[12] led to renewed interest in the area. In addition, interest in reliability and failure-time data in industrial statistics[13] led to the examination of other parametric models for describing time to response data. A diversity of methods for examining these types of data exist: parametric models (classical distribution fitting) through semi-parametric models (proportional hazard models are sometimes referred to in this way) to non-parametric models (percentile estimation and smoothing). There are several reasons for this plurality, but, some instances, relate to the differences in the nature and objectives of experiments and to the fact that different methods have developed in different application areas. Two useful general references are Collett[14] and

Oakes.[15] In the context of insect pathology, no standard protocol exists for experiments with different numbers of doses, subjects and different observations figuring in different assays. There is perhaps an over-emphasis on comparisons of LT50s, but the major objective is certainly to discriminate between doses on the basis of response time.

7.4 Insect pathology

Insect pathology is the study of diseases in insects, but has become very much associated with the control of insect pests using pathogenic organisms. A useful review of the field can be found in Payne.[4] Pathogenic microorganisms include viruses, bacteria and fungi. Viruses and bacteria are usually ingested by the insect host; in the case of viruses they replicate within the host, eventually killing them, while bacteria operate by releasing a toxin that essentially paralyzes its victim. Fungal pathogens attack their hosts via spores that penetrate the cuticle of the insect, producing a mycelium that spreads through its body. Another control agent that is often grouped with the above are entomopathogenic nematodes — microscopic worms that invade insects and release an infectious bacterium.

Many of the studies involving these pathogens have been more concerned with time to response and impact rather than lethal dose. Historically, research has often centered on comparisons between, and development of, new strains in an attempt to shorten the time to response. The fact that biopesticides kill more slowly than chemical pesticides has always put them at a commercial disadvantage, although "green accounting" and consumer perception are making them more acceptable despite their response deficiencies. In the last few years, these studies have been given greater impetus by the development of new strains through bioengineering.

In the development that follows, we will look at the types of data that such studies produce and consider how these data sets can be analyzed.

7.4.1 Experimental procedures

Details of experimental protocols can be found in standard journals of insect pathology such as the *Journal of Insect Pathology* and *BioControl Science and Technology*. Generally, an experiment will consist of several groups of insects (of the order of six groups of 25 or so insects) that are exposed to the pathogen for a period of 24 hours. A control group is usually included (this group will be treated identically to the experimental groups, except there will be no pathogen present) to ensure that mortality does not occur through handling, etc. Various checks are done to ensure that the insect has ingested the pathogen, and the test hosts are then moved to fresh, uncontaminated diet and observed regularly (e.g., daily, or possibly more frequently) until the cumulative mortality has stabilized. Insects are usually housed individually, often in micro-titer wells, though communal housing also occurs. Later, we will consider an example of the latter that can lead to unusual results.

7.4.2 Basic data analysis

Data are often collected on a cumulative basis for each dose, and monitoring ceases when the numbers of dead insects have stabilized. The first point to appreciate is that these data are highly correlated when presented in cumulative form, since the count on day $t + 1$ is very dependent on that of the previous day, t. While it is very informative to look at the accumulated count over the course of the experiment, this will not be the natural way to analyze the data. Table 7.1 shows the results of a typical assay, which are represented as a cumulative mortality plot in Figure 7.1. A second feature of the data is

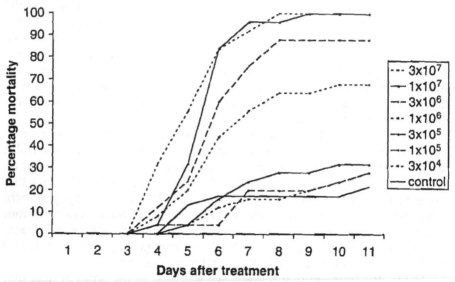

Figure 7.1 Plot of bioassay results in Table 7.1

Table 7.1 Results from a Bioassay Investigating the Mortality of *Plutella xylostella* Larvae Offered Different Concentrations of Px Granulosis Virus

Conc. (OB ml⁻¹)	No. of larvae	No. of days after treatment								
		3	4	5	6	7	8	9	10	11
control	25	0	0	3	4	4	4	4	4	5
3×10^4	25	0	1	1	3	4	4	5	6	7
1×10^5	25	0	1	1	1	5	5	5	6	7
3×10^5	25	0	0	1	4	6	7	7	8	8
1×10^6	25	0	2	5	11	14	16	16	17	17
3×10^6	25	0	3	6	15	19	22	22	22	22
1×10^7	25	0	1	8	21	24	24	25	25	25
3×10^7	25	0	8	14	21	23	25	25	25	25

The data show the cumulative mortality of insects over time.

that they are censored, i.e., we do not know the exact time to death of each individual (interval censoring), nor are we sure that all treatment-induced mortality has fully occurred at the time that sampling ceases (right-censoring). Both types of censoring affect the assumptions we can make about the data and the way in which they can be analyzed. Note also in the above data set that control mortality is non-zero after day 5, and this must affect the interpretation of the mortality of treated insects. Before the time response was seen to be important, assays of this type would often be analyzed using probit analysis for the last time sample in much the same way as a probit analysis for a conventional pesticide. Such an approach is practically identical to that adopted in conventional ecotoxicological testing.[16]

7.4.3   *Preliminary analyses*

Although data from these studies are typically presented in the manner of Table 7.1 (to correspond to the cumulative mortality curves of Figure 7.1) a truer representation is given in Table 7.2. This shows the mortality on a daily basis, and describes the basic distribution of time to death, except, for the lower doses in particular, a substantial proportion of the insects have not responded by the end of the measurement period. However, Figure 7.1 suggests that, by day 10, most of the virus-associated mortality should have occurred. To take account of the dichotomy between survivors and victims we must utilize the two independent pieces of information associated with them: first, estimate the proportion of insects dying in the experimental period; and then, separately, consider simple statistical summaries of the victims' time to death. These are given in Table 7.3.

Recent work has questioned how the control group should be treated in this type of study,[17,18] particularly as it is usually possible to distinguish between insects dying from viral infection and those dying from other causes. It is worth noting that the control mortality is quite large in this

Table 7.2 Data from Table 7.1 Reorganized to Show the Number of Larvae Dying on Different Days

Conc. (OB ml⁻¹)	No. of larvae	No. of days after treatment								
		3	4	5	6	7	8	9	10	11
control	25	0	0	3	1	0	0	0	0	1
3×10^4	25	0	1	0	2	1	0	1	1	1
1×10^5	25	0	1	0	0	4	0	0	1	1
3×10^5	25	0	0	1	3	2	1	0	1	0
1×10^6	25	0	2	3	6	3	2	0	1	0
3×10^6	25	0	3	3	9	10	3	0	0	0
1×10^7	25	0	1	7	13	1	0	1	0	0
3×10^7	25	0	8	6	7	2	2	0	0	0

Table 7.3 Summary Statistics of Time to Death

log₁₀ dose	Proportional mortality	Mean time to death	s.d.
Control	0.22	6.4	2.61
4.48	0.28	7.6	2.51
5.00	0.28	7.6	2.30
5.48	0.32	6.9	1.55
6.00	0.68	6.2	1.52
6.48	0.88	6.0	1.21
7.00	1.00	5.9	0.97
7.48	1.00	5.4	1.25

example, and that mortality on day 5 is consistent with viral death, which might suggest viral contamination of the control group.[19] Certainly, Figure 7.1 suggests that the control group response is reasonably consistent with that for low-dose treatments. It is clear from the data summary in Table 7.3 that the mean time to response *and* its standard deviation decline with increasing dose. This suggests that, not only does the higher dose hasten the time to death, but also induces a faster response once the process has started. In effect, three aspects of the original data are important and they correspond to the three parameters or summary statistics in Table 7.3:

1. The proportion of insects that respond at each dose
2. The mean time to death of the respondents
3. The standard deviation of the time to death

The above is very simplistic, but reasonably robust in the context of such data. A more sophisticated way of analyzing such data is to use a maximum likelihood approach. Hunter et al.[20] show how this is done for normally distributed data. The method takes full account of the interval censoring of the data by determining the probability of death between consecutive days and matching these to the data, the multinomial distribution being used as a device for equating actual and expected numbers.

Typically, one might take the data in Table 7.3 and examine the three parameters separately with respect to the dose parameter. In fact, for (1) this would simply be the conventional quantal assay for day 11. Then we might wish to look at the changes in mean (2) and standard deviation (3) for time to death at different doses, and maybe model them. These are presented as simple plots in Figure 7.2, and we do not take (2) and (3) any further. However, it is possible to look at the whole process in an integrated way, and Morgan[5] explores such models in detail. Terrill[17] considers a general time to response model for the above data.

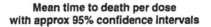

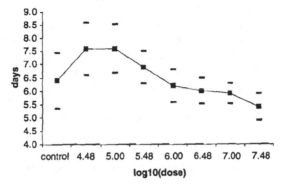

Figure 7.2 Mean time to death and proportion responding.

7.4.4   *Modeling each sample point*

An alternative way of looking at the data is also instructive, although, strictly speaking, less correct. Earlier, it was pointed out that a probit analysis of responses at the termination of the assay corresponds to what typically occurs in toxicity testing. If this analysis is extended backward through time we can derive a profile of how the LD50 is changing over time. The model we have used is actually a logit-based analysis with estimated control mortality, and Figure 7.3 shows the changing LD50 with time, together with the upper and lower 95% fiducial limits. Superimposed on the figure are fitted exponential decay responses to these three statistics, which, it can be seen, fit almost perfectly. From these, we have asymptotic estimates of the LD50 and the corresponding fiducial limits, which are, respectively, 5.94 and (5.58; 6.16). The residual standard deviation from fitting is of the order of 0.02, which is very small. It is instructive to note that the actual LD50 is closer to the upper than the lower limit, which one might expect. Referring back to Figure 7.1, this is essentially because the high control mortality forces the LD50 closer to the dose of 10^6 OBS ml^{-1} and the doses above that value give more information than those below, all of which are close to the control mortality. In Figure 7.3, we also see how estimated control mortality changes steadily over time despite the fact that the actual control mortality is stable between days 6 and 10. When control mortality is estimated, the estimate also takes information from the low doses when they have responses close to that of the control group. So, while the control group itself may be stable for a period, the response to lower doses is creeping up, which is reflected in the control mortality estimate. The relatively high levels of control mortality on days 5 and 6 (higher than three lower doses of the trial) might lead to a rejection of thecontrol mortality data as being atypical. This is a real problem in many insect pathology experiments where the target insect is

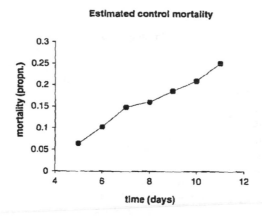

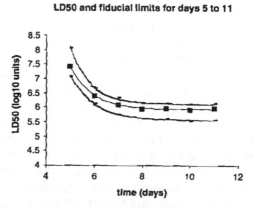

Figure 7.3 LD50, fiducial limits and control mortality.

fragile, and "natural" or control mortality is to be expected. A possible solution is to increase the size of each experimental group.

7.4.5   Other probability distributions

In discussing the problem so far, we have considered only the normal distribution, but this is only one among a whole spectrum of distributions that could be used to describe survival data. The use of many of these distributions has arisen in reliability theory, particularly with reference to industrial statistics.[13] The distinction between them tends to lie mainly in the extended right-hand tail; these distributions are almost exclusively skewed, unlike the normal distribution. Extensive data are usually required to distinguish between them, a luxury insect pathologists may not have, because the target host may well metamorphose prior to complete data collection. Indeed it has been observed (Rovesti, personal communication) that sublethal doses of virus can interact with the life cycle of an insect. Nevertheless, it is important to be aware of the distinction between symmetric distributions such as the normal and logistic, and their asymmetric logarithmic counterparts and other skewed distributions such as the Weibull and gamma. Kent and Queensberry[21] give a comparison of the exponential, lognormal, gamma and Weibull distributions. See Morgan[5] for other potential distributions in this area, though experiment size in the field of insect pathology is often insufficient to discriminate between similarly shaped distributions.

In an early paper, Bliss[5] devoted considerable time to the graphical analysis of data. For the most part, he considered probit-type responses, i.e., the normal distribution on a logarithmic time-scale, that, he considered, best fitted his data. Some of this obviously derived from the empirical finding that response to dose seemed approximately normal on the logarithmic scale. Nevertheless, Bliss's examples were consistent and he demonstrated how a plot of probit versus log (time) gave an excellent straight line, whereas the same plot using untransformed time or reciprocal time produced curves that were, respectively, convex upward or downward. There is a simple relationship between different distributions and the corresponding transformations to achieve linearity:

Survival distribution	Cumulative distribution	Transformation
Normal (Probit)	$\phi(a+bt)$	$\phi^{-1}(p)$
Logistic	$[1+1/\exp(a+bt)]^{-1}$	$\log[p/(1-p)]$
Weibull	$1-\exp[-\exp(a+bt)]$	$\log[-\log(1-p)]$

For practical purposes, the substitution of the cumulative proportion of insects surviving (or dying) into the transformation in column 3, and the corresponding plot against time, or log(time) should produce a straight line. A difficulty arises, although it is usually ignored, in that the general methods of estimating parameters for the fitted model rely on an assumption about the underlying distribution. If we use the data to search for the right distribution by attempting to find a linear plot, we are effectively re-using the data, first to find the distribution and second to estimate the parameters. However, one can claim, as did Bliss, that a particular type of distribution appears to better fit experimental data from a certain class, e.g., biological control of entomophagous pests, and that the plots are simply to check that models for a series of assays with different doses are consistent. This can be most easily checked by logits or probits versus time or log (time) for all assays on the same frame.

7.4.6 Survival analysis

Another approach to this type of data is to use methods commonly referred to as survival analysis methods.[5,22] This is usually effected by modeling a hazard function, which can be thought of as the instantaneous death rate. Suppose that the probability that an insect dies in the interval $(t, t + \delta t)$ is expressed as $\lambda(t)\delta t$, then $\lambda(t)$ is the *hazard function*. The survival times T of individuals in a population are assumed to have a density function $f(t)$. The corresponding distribution function $F(t)$ is given by

$$\lambda(t) = \frac{f(t)}{1 - F(t)} \qquad\qquad 7.1$$

which can be rearranged as

$$F(t) = 1 - \exp\left\{-\int_{-\infty}^{t} \lambda(s)ds\right\} \qquad\qquad 7.2$$

In the context of insect pathology we might choose to think of $\lambda_0(t)$ as representing a base-line hazard — this could be the control group or a low-dose group — and then we frequently write

$$\lambda_i(t) = \lambda_0(t)k_i(t) \qquad\qquad i = 1, 2 \dots p \qquad\qquad 7.3$$

This is known as a *proportional hazards model*, as the hazard for group i is $k_i(t)$ times that of group 0; $k_i(t)$ often takes a simple exponential form such as $\exp(\beta_i + \gamma_i(t))$.

An excellent description of these models in an entomological context is given in Rowell and Markham.[23] In particular, these authors use a "semi-parametric" model in which the hazard function is allowed to change in a non-smooth way, unlike that for parametric models. Rather than attempting to fit a response to the cumulative mortality data, the model considers the probability of death in an interval, conditional on the number of insects alive at the start of the interval. To illustrate this, consider the data in Table 7.4.

Table 7.4 Data from a Second Assay Showing the Cumulative
Mortality of *P. xylostella* Challenged with *PxGV*

Conc.	No. of	No. of days after treatment					
(OB ml⁻¹)	larvae	3	4	5	6	7	8
1×10^5	32	0	0	3	8	8	16
3×10^5	28	0	0	1	9	10	11
6×10^5	35	0	0	7	12	12	24
1×10^6	28	0	0	7	18	20	25
3×10^6	27	0	0	9	17	18	20
6×10^6	34	0	0	13	30	30	31
1×10^7	37	0	8	23	35	35	36
3×10^7	37	0	6	33	33	36	37
6×10^7	34	0	3	21	34	34	34

These data are taken from another *PxGV* assay and illustrate an interesting phenomenon: the hiatus in response between days 6 and 7, which is common to nearly all doses. For the higher doses, mortality has already reached 90% so that lack of response is not very noticeable, but, at lower rates, the cumulative mortality shows a dramatic upturn on the following day. This assay was performed on groups of insects, so that here, secondary infection from organisms released from cadavers might be the cause of the upsurge in response. The "flat" response between days 6 and 7 reflects the likely true mortality caused by the original dose. Simple logit analysis of the data from

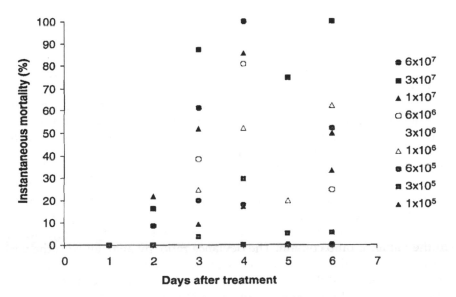

Figure 7.4 Instantaneous death rates for the data in Table 7.4.

these two days shows that the LD50 for day 6 is 5.78 ($\log_{10}$) with 95% fiducial limits of 5.59–6.09, while that for day 7 is 5.80 with corresponding limits of 5.56–6.00, i.e., there is no discernible difference.

To model these data, we form the empirical hazard function γ_{it} from the mortality data of Table 7.4; the suffixes i and t correspond to dose and time respectively. Recalling that the hazard function is the instantaneous death rate, we can calculate it from the number of insects dying in a particular period relative to the number alive at the start of the period. Thus

$$\lambda_{it} = \frac{r_{i,t+1} - r_{it}}{n_{i0} - r_{it}}$$

7.4

where n_{i0} is the initial number of insects at time $t = 0$ for treatment i. These values are plotted in Figure 7.4, which suggests that the instantaneous death rates are quite consistent over the duration of the experiment, in that the rank order of dose corresponds to the instantaneous hazard or risk. Using a generalized linear model approach (as described in Rowell and Markham[23]) a hierarchy of models was fitted to these data: equal hazards (in which the basic hypothesis is that there is no difference between doses), proportional hazards (where hazards are linked by a proportionality function that might vary over time, as described at the beginning of this section) and unequal hazards (in which all time × dose parameters are essentially independent.) The idea is analogous to that of analysis of variance in that the day to day variation in average rates is similar to block to block variation, with proportional hazards seeking a consistent proportional response between doses. The corresponding disproportionate hazards model represents the case in which hazard rate varies from treatment to treatment across time. The outcome of the analysis is shown in the top half of Table 7.5, where it can be seen that the proportional hazards model (Model B) provides a good description of the data, i.e., it is significantly better than the equal hazards model (Model A), and it is not bettered by the disproportionate hazards model (Model C).

What we are effectively doing in exploring this hierarchical model structure is looking for the most parsimonious description of the data, i.e., the one with the least parameters that still accounts for much of the explainable variation. In interpreting the table, we first move to the most complex model (sometimes called the saturated model), which, in this example, is Model C, and look to see if the deviance (note that this is also the model with the smallest deviance) is consistent with the χ^2 distribution. If it is, then the various tests are conducted on the deviance column using the χ^2 distribution, otherwise, an approximate F test is performed on the corresponding statistics in the variance ratio column. The deviance ratio, by analogy with the variance ratio in analysis of variance, is the ratio of the mean deviance for a particular effect and the mean deviance (effectively the residual mean deviance) from the saturated model. To compare models, we look at the change

Table 7.5 Analysis of Deviance Table for Proportional Hazards Model and Some Extensions

Model	Source	df	Deviance	Mean deviance	Deviance ratio	
A	Equal hazards	40	148.9			
B	Proportional hazards	32	69.06			
C	Unequal hazards	24	46.44	1.94		
	A-B	8	88.8	11.1	5.74	p < 0.001
	B-C	8	13.62	1.70	0.88	n.s.
D	1/dose	39	72.98			
	A-D	1	75.9	75.9	39.12	p < 0.001
	D-C	15	26.54	1.77	0.91	n.s.

in deviance in moving from one model to another, so that in testing to see if a proportional model is appropriate, we consider the difference "B-C" (row 5), and find that the mean deviance for the change is, in fact, smaller than the residual mean deviance (that for Model C). We then test to see if there is evidence for the proportional hazards model compared with an equal hazards (or no treatment difference) model. Clearly, from the significant ratio obtained in row 4 (labeled "A-B") there is considerable evidence for the proportional hazards model ($p < 0.001$).

The parameter estimates in the proportional hazards model are not directly related to the underlying dose, and it would be useful if the relationship between doses could be described in a simple quantitative way. This is now done by using a simple inverse function to describe the form of the relationship between the hazard function $\lambda_i(t)$ for dose i, and that for the lowest dose, $\lambda_0(t)$, in the very simple form:

$$\lambda_i(t) = \lambda_0(t).\exp(-\log(dose))$$
$$= \lambda_0(t)/dose$$

7.5

This is Model D, and Table 7.5 shows that, although the model has only one more parameter than the (unrealistic) equal hazards model, that single parameter that defines the proportionality in terms of dose is as good a representation of the data as the proportional hazards model with seven more parameters (one for each individual dose tested).

7.5 Discussion

This chapter has attempted to show how time to response data is sometimes analyzed in the field of insect pathology. The emphasis has been practical rather than mathematical, although a limited mathematical development was introduced in the survival analysis section. Two multiple dose examples have been used to illustrate the basic methods. It is possible to consider all the models presented here in a general time to response framework,[17] but we have considered a more simple, summary statistics approach in the first example, which also shows how end-point analysis fails to capture the considerable amount of information contained in the repeated measure data prior to the end point. In the second example, survival analysis methods are used and the example demonstrates how a model incorporating both dose and time can be developed.

In the field of insect pathology, the scientist is sometimes comparing different organisms, but a major element of the type of data described here is simply to gain an understanding of the interaction of dose and time on the host insect. This often means no obvious testing framework for the data, and statistical modeling methods are used to provide parameter estimates and standard errors that give information on the impact of different dose levels. In the context of regulatory ecotoxicology, such an approach could prove valuable, particularly if it is coupled to a formal decision analysis framework.[24] Alternatively, a model could be used to estimate the response to different dose (or toxin) levels. Whichever approach is adopted, considerable attention will need to be paid to the experimental protocol.

The experimental procedures described for insect pathology are not a standard but rather a general protocol, for, in experimental science, the aims of individual experiments vary. For environmental toxicology, it will be necessary to establish formal protocols that will enable the objectives of the regulatory authorities to be realized. In particular, robust experimental designs with standard protocols for replication, sampling, etc., will need to be established, and a methodology for analysis laid down. It is also important that various diagnostic methods be incorporated into any protocol to ensure that the assumptions of the statistical analysis methods are valid. One area that will need careful monitoring is the level of control (or natural) mortality that was present in the examples considered here, but was not examined in detail.

7.6 Conclusions

The methods described above show how time to response data (for quantal, or simple two-state outcomes) might be examined in a variety of ways. The examples demonstrate that both end-point response and the distribution of mortalities conditional on it can vary. While the emphasis has been more on modeling than on routine analysis, frequently a requirement for comparative studies, as well as for quality assurance, exists. What is important in all aspects of scientific experimentation is that the objectives of any assay be clear. If these methods are established in environmental toxicology, it will be necessary to formulate protocols based on robust experimental designs that can provide sufficient discrimination for the objectives laid down by regulatory authorities. Furthermore, it is important that diagnostic methods (e.g., an examination of the goodness of fit of any model to the data) are incorporated in any protocol to ensure that the assumptions of the analytical methods are valid. But there can be little doubt that intermediate responses of time-course studies, prior to the endpoint, provide valuable additional information in informing decisions.

References

1. Finch-Savage, W.E., Steckel, J.R.A. and Phelps, K., Germination and post-germination growth to carrot seedling emergence: predictive threshold models and sources of variation between sowing occasion, *New Phytologist*, 139, 505, 1998.
2. Aikman, D.P. and Scaife, A.J., Modeling plant growth under varying environmental conditions in a uniform canopy, *Ann. Bot.*, 72, 485, 1993.
3. Collier, R.H. and Finch, S., Field and laboratory studies on the effects of temperature on the development of the carrot fly (*Psila rosae* F.), *Ann. Appl. Biol.*, 128, 1, 1996.
4. Payne, C.C., Pathogens for the control of insects: where next?, *Philos. Trans. Royal Soc.*, Series B, 318, 225, 1988.
5. Morgan, B.J.T., *Analysis of Quantal Response Data*, Chapman & Hall, London, UK, 1992.
6. Bliss, C.I., The calculation of the dosage-mortality curve, *Ann. Appl. Biol.*, 22, 134, 1935.
7. Sampford, M.R., The estimation of response-time distribution. I. Fundamental concepts and general methods, *Biometrics*, 8, 13, 1952.
8. Sampford, M.R., The estimation of response-time distribution. II. Multi-stimulus distribution, *Biometrics*, 8, 307, 1952.
9. Sampford, M.R., The estimation of response-time distribution. III. Truncation and survival, *Biometrics*, 10, 531, 1954.
10. Litchfield, J.T., Jr., A method for rapid graphic solution of time-percent effect curves, *J. Pharmacol. & Exp. Ther.*, 97, 399, 1949.
11. Nelder, J.A. and Wedderburn, R.W.M., Generalized linear models, *J.Royal Stat. Soc.*, Series A, 135, 370, 1972.
12. Cox, D.R., Regression models and life tables, *J.Royal Stat. Soc.*, Series B, 34, 187, 1972.
13. Crowder, M.J., Kimber, A., Sweeting, T. and Smith, R., *Statistical Analysis of Reliability Data*, Chapman & Hall, London, 1994.
14. Collett, D., *Modelling Binary Data*, Chapman & Hall, London, 1991.
15. Oakes, D., Life-Table Analysis, in *Statistical Theory and Modelling*, Hinkley, D.V., Reid, N. and Snell, E.J., Eds., Chapman & Hall, London, UK, 1991, 107.
16. Pack, S., *A Review of Statistical Data Analysis and Experimental Design in OECD Aquatic Toxicology Test Guidelines*, Shell Research Ltd., Sittingbourne Research Centre, Sittingbourne, UK, 1993.
17. Terrill, P.K., *Statistical Models in the Assessment of Biological Control of Insects*, PhD thesis, University of Kent at Canterbury, 1997.

18. Long, S.J., Richardson, P.N. and Fenlon, J.S., Influence of temperature on the infectivity of entomopathogenic nematodes (*Steinernema* and *Heterorhabditis* spp.) to larvae and pupae of the vine weevil *Otiorhynchus sulcatus* (Coleoptera: Curculionidae), *Nematology*, 2, 309, 2000.

19. Fenlon, J.S., How can a statistician improve your interpretation?, *Pest. Sci.*, 45, 77, 195.

20. Hunter, E.A., Glasbey, C.A. and Naylor, R.E.L., The analysis of data from germination tests, *J. Ag. Sci.*, Cambridge, 102, 207, 1984.

21. Kent, C.P. and Queensberry, J., Selecting among probability distributions used in reliability, *Technometrics*, 24, 59, 1982.

22. Cox, D.R. and Oakes, D., *Analysis of Survival Data*, Chapman & Hall, London, UK, 1984.

23. Rowell, J.G. and Markham, P.G., Modeling mortality risks for treated insects relative to controls, *Ann. Appl. Biol.* 105, 15, 1984.

24. Smith, J.Q., *Decision Analysis: a Bayesian Approach*, 2nd ed., Chapman & Hall, London, UK, 1997.

chapter 8

Time to event analysis in ecology

Bryan F.J. Manly

Contents

8.1 Introduction

Time to event studies in ecology are reviewed in this chapter, and three examples that involve the foraging time of Antarctic penguins, the survival of fish in water contaminated with metals, and the survival time of goslings in nests in Alaska are described. These examples illustrate the unusual types of data that can occur in ecology. In discussing the analysis of the data, emphasis is given to the use of the computer-intensive methods of randomization and bootstrapping as useful alternatives to more conventional approaches for testing hypotheses and assessing the accuracy of estimates.

8.2   Time to event in ecology

The analysis of time to event data is of interest in ecology in a variety of situations. Some examples of such data are the length of life of an animal or plant, the duration of a developmental stage in a life cycle, the duration of an activity such as foraging by an animal, length of time that an animal is in a study area, and the time to extinction of an endangered species. Sometimes the variable of interest is observed directly on a sample of individuals and standard statistical methods for analyzing survival data can be applied.[1,2] However, some situations where the nature of the data makes the analysis more complicated have necessitated development of special methods.

An example of an application of special interest is the determination of the distribution of the time to extinction of an endangered species, an area that is sometimes called population viability analysis. Typically, this involves setting up a stochastic population model using whatever demographic information is known about the species, and then simulating possible future trajectories for the population to determine the probability of extinction after 20 years, 50 years, 100 years, and so on.[3] Special computer programs are available to facilitate this process.[4,5] Two other areas where specialized methods have been developed are the estimation of the distribution of the durations of developmental stages for animals and plants from laboratory and field data,[6] and the estimation of the survival rates of mobile animals by marking and recapture.[7] Again, special computer programs are available to do the calculations.[8,9]

The remainder of this chapter is concerned with three particular examples that are used as a framework for discussing methods of analysis, with an emphasis on the use of randomization and bootstrapping. These approaches to data analysis have received increasing use in recent years[10] because they tend to require fewer assumptions than alternative more conventional methods, and inferences can often be based on quite simple principles that are easy to understand.

The first example concerns the duration of the foraging time (the time between leaving and returning to a nest) of Adélie penguins (*Pygoscelis adeliae*) in the Antarctic. This variable is one of many used to monitor the health of the Antarctic ecosystem by the Working Group on Ecosystem Monitoring and Management (WG-EMM) of the Commission for the Conservation of Antarctic Marine Living Resources (CCAMLR). It has a highly non-normal distribution within colonies, and the sampling methods that are used result in data with a complicated correlation structure. These factors mean that testing for significant differences between mean foraging times in different years, for different colonies, for different stages (guarding the eggs and guarding the chicks), and for different groups of penguins in one colony at one time is not straightforward. Nevertheless, these tests are important for the CCAMLR monitoring program.

With this example, randomization can be used to test for differences between mean foraging times for male and female penguins, and bootstrapping can be used to determine the standard errors associated with estimated mean durations. The bootstrap standard errors can then be used as weights for an analysis of variance to compare mean foraging durations under different conditions. The example shows how randomization and bootstrapping can, with reasonable ease, accommodate a complicated data structure that is difficult to handle using more conventional parametric methods.

The second example concerns part of the results from a series of experiments conducted by Marr et al.[11] to compare the survival of naive and metals-acclimated juvenile brown trout (*Salmo trutta*) and rainbow trout (*Oncorhynchus mykiss*) when exposed to a metals mixture with the maximum concentrations found in the Clark Fork River, Montana. The experiment considered here (challenge 1) followed three groups of fish (hatchery brown trout, Clark Fork River brown trout, and hatchery rainbow trout). Approximately half of each group (randomly selected) were controls that were kept in clean water for 3 weeks before being transferred to the metals mixture. The rest of the fish in each group were acclimated for 3 weeks in a weak solution of metals before being transferred to the stronger mixture. All fish survived the initial 3-week period, and the outcome variable was the survival time of the fish in the stronger mixture.

Marr et al.[11] analyzed mean survival times using analysis of variance with the weight of fish as a covariate and using a series of t-tests to compare pairs of treatments. Here, two alternative analyses are discussed. The first uses multiple regression with randomization inference. This is essentially similar to the analysis of variance used by Marr et al. except that it was conducted on logarithms of survival time to stabilize the error variance, which otherwise appears to increase with the mean survival time. The second alternative analysis uses restricted randomization, and is less model-based. Like the randomization tests used with the penguin data, this example demonstrates how methods of randomization can be designed specifically for the data at hand and the null hypotheses of interest.

The third example concerns the survival time of goslings of emperor geese (*Chen canagica*) on the Yukon-Kuskokwim Delta, Alaska, in 1993 and 1994. For both years, the number of goslings either at hatching or some later time is known for a number of broods, and also the numbers still surviving at various later times up to about 40 days. There is interest in estimating the proportion of goslings in the population surviving different amounts of time, and also whether this survival function varies in different years. However, there are three aspects of the situation that make an analysis somewhat complicated. First, the observation times are quite irregular. Second, it is possible that survival probabilities will vary from brood to brood. Third, it sometimes happens that a gosling will transfer from one brood to another so that the number observed in a brood can increase from one observation time to the next. This third complication means that the usual distribution theory for standard parametric models for survival data cannot be applied because, according to this theory, some of the observations are impossible.

To cope with this situation, a survival function for a year can be estimated by nonlinear least-squares, and bootstrap resampling of broods can be used to assess the accuracy of the estimates obtained. A randomization test for differences between the survival in two or more years is also possible.

8.3   Analysis of Adélie Penguin data

8.3.1   Penguin ecology

Adélie penguins live and breed around the entire Antarctic continent. Their foraging ranges during the chick rearing period vary between locations, with those breeding on the Antarctic Peninsula traveling relatively short distances (14–60 km) to feed mainly on krill, and those breeding on the continent tending to travel farther (up to 110 km) to feed on a variety of organisms from over the Antarctic shelf, as well as in deep water.[12]

The data considered here were collected at two locations. The first is Bechervaise Island, which is a long-term monitoring site under the CCAMLR Ecosystem Monitoring Program. There, research on Adélie penguins has been funded by the Australian government on a colony of about 1800 breeding pairs since 1990. The second location is Edmonson Point, where a CCAMLR monitoring site was established during the 1994–95 summer to allow a comparison with Bechervaise Island for an Adélie penguin colony of about the same size. The two sites differ because Bechervaise Island penguins have access to deep water off the edge of the continental shelf, while those at Edmonson Point forage exclusively in shallower waters.

Foraging duration is monitored as part of the CCAMLR program because this is a measure of the availability of food to the penguins. It is presumed that difficulty in finding food will be reflected in an increase in the foraging duration. Details of recording methods are described by Clarke et al.[12] Here it is only necessary to note that foraging-trip durations of male and female breeding birds were recorded for 150 marked nests at Bechervaise Island in 1991–92, 1992–93, 1993–94 and 1994–95 and at Edmonson Point in 1995–96 for both the guard stage when eggs were present, and the crèche stage when chicks were present. In addition, foraging durations were recorded for 120 nests over a 10-day period during the guard stage at Edmonson Point in 1994–95.

8.3.2   Statistical analysis

With these data, there is interest in comparing the mean foraging durations in the two colonies in the different seasons, and also between the guard and crèche stages. In addition, Clarke et al.[12] were particularly interested in any differences in the mean duration between male and female penguins because it has, until now, been assumed that the two sexes have similar foraging strategies because they show little sexual dimorphism.

Consider first the question of whether there is a significant difference between the male and female mean duration for one of the two stages (guard and crèche) at a particular site in a particular season. This comparison is complicated by two considerations. First, the number of measured foraging durations varies from penguin to penguin. For example, for the crèche stage at Edmonson Point in 1995–96, there were from one to 21 durations recorded for individual penguins. Second, there may be variation associated with individual nests, but there are some records for just the male or just the female in a nest, and other records for both.

A randomization test can accommodate these complications in a relatively straightforward way. The null hypothesis is that the labels "male" and "female" on the penguins are unrelated to foraging durations, so that any difference between the male and female mean could have arisen by a random allocation of these labels. The procedure for this test is as follows:

1.   The mean duration for all males is calculated, giving equal weight to all observations irrespective of how many are available for each penguin. Similarly, the mean is calculated for all females. The difference D_1 (female–male) is then calculated.

2.   A randomized set of data is produced in two stages. First, the nests with records for one penguin (male or female) are considered. The labels "male" and "female" are randomly allocated to the penguins in these nests, keeping the total number of each sex the same as in the original data. Next, the nests with both a male and a female are considered. The labels "male" and "female" are switched with probability 0.5.

3.   Mean values and their difference (D_2) are calculated for the randomized data, as described in step 1.

4.   Steps 2 and 3 are repeated 4999 times to generate randomized differences $D_2, D_3, ..., D_{5000}$. The original difference D_1 is then significantly different from 0 on a two-sided test if it is either one of the largest 2.5% or one of the smallest 2.5% of the full set of 5000 differences. If there is no male–female difference, the probability of this happening by chance is 0.05, as required. Similarly, D_1 is significantly different from 0 at the 1% level if it is among the largest 0.5% or the smallest 0.5% of the 5000 differences, or significantly different from 0 at the 0.1% level if it is among the largest or smallest 0.05% of the 5000 differences.

At step 1 it is possible to calculate the mean foraging duration for each individual penguin, find the average of these means for males and females, and then use the difference between these averages as the test statistic. This was found to give essentially the same results as using the mean difference as defined for step 1.

A two-step approach was used to assess the differences between mean foraging durations associated with different colonies, seasons and stages. First, standard errors were estimated for mean values using a bootstrap method. Then, analysis of variance was used to assess the significance of factors and interactions using the bootstrap standard errors for weights.

The bootstrap procedure for estimating standard errors was carried out separately for the two sites, for each of the stage-season-year combinations. The procedure was as follows:

1. Consider the individual penguins (males and females) for which foraging times are available but where there is no information for the other penguin in the nest. This set of penguins is treated as a "population" and sampled with replacement to obtain a new (bootstrap) sample of penguins. Each of the penguins in this sample has a certain number of foraging times recorded. These are sampled with replacement to get new times for that penguin. In this way, a bootstrap set of data is generated for single penguins, taking into account both variation among penguins and variation between trips for one penguin. In all cases, a resampled penguin keeps its original sex. The number of recorded trips for each penguin and the number of penguins are also kept constant.

2. Consider the records for pairs of penguins in nests. These are treated as a "population" that is resampled with replacement. Each penguin in each chosen nest is then given a new set of foraging times by resampling from the ones that are available for it. In this way, new bootstrapped pair data are generated. The number of nests, the sex of penguins, and the number of trips per penguin are all kept equal to the values for the real data.

3. Mean foraging times and the female–male mean difference were determined from the bootstrap data.

4. Steps 1 to 3 are repeated 5000 times. The standard error of any parameter such as the female mean duration is estimated by the standard deviation of the estimates calculated from the bootstrapped data.

In using this procedure, it is being assumed that the foraging times observed for a single penguin are a random sample from a constant distribution, which, from an inspection of the data, appears to be reasonable.

8.3.3 Results

The results obtained from the randomization tests and bootstrapping are shown in Table 8.1. The chicks all died at Bechervaise Island in 1994-95 so that mean foraging durations cannot be calculated for the crèche stage in

Table 8.1 Mean Foraging Trip Durations for Adélie Penguins, with Standard Errors Determined by Bootstrapping, and the Results of Randomization Tests for the Difference Between Males and Females

Season	Sex	Mean duration(h)	SE	Female–Male difference	SE	Test result[1]	Sample size (trips)	Sample size (birds)
		Guard Stage, Bechervaise Island						
91–92	m	26.1	2.9	9.4	3.9	**	344	57
	f	35.5	2.8				449	63
92–93	m	32.5	3.2	12.0	3.9	***	371	74
	f	44.5	2.4				380	71
93–94	m	24.8	1.3	10.0	2.1	***	290	71
	f	34.8	1.7				296	77
94–95	m	36.2	4.6	37.2	8.2	***	198	67
	f	73.3	7.9				209	76
		Guard Stage, Edmonson Point						
94–95	m	26.0	0.9	5.5	1.3	***	260	94
	f	33.5	1.3				255	97
95–96	m	25.3	0.9	5.6	1.5	***	235	55
	f	30.9	1.2				288	61
		Crèche Stage, Bechervaise Island						
91–92	m	26.4	1.7	5.0	2.4	*	590	46
	f	31.4	1.9				588	52
92–93	m	39.4	2.3	5.1	3.2	*	637	71
	f	44.5	2.4				566	66

continued

Table 8.1 (continued) Mean Foraging Trip Durations for Adélie Penguins, with Standard Errors Determined by Bootstrapping, and the Results of Randomization Tests for the Difference Between Males and Females

Season	Sex	Mean duration(h)	SE	Female–Male difference	SE	Test result[1]	Sample size (trips)	Sample size (birds)
93–94	m	34.1	1.4	3.6	2.4	NS	961	77
	f	37.7	1.9				884	76
94–95	m			Chicks dead				
	f			Chicks dead				
Crèche Stage, Edmonson Point								
94–95	m			No records available				
	f			No records available				
95–96	m	15.7	0.9	6.2	1.9	***	329	35
	f	21.8	1.7				199	35

[1]For randomization tests significance levels are: NS, not significant at the 5% level; *, significant at the 5% level; **, significant at the 1% level; and ***, significant at the 0.1% level.

this case. The mean foraging duration was always higher for females than it was for males, with all except one of the differences being significant at the 5% level at least.

To examine whether the difference between male and female foraging durations varies with other factors, an analysis of variance was carried out using the computer package GLIM,[13] with the individual female–male means weighted using the reciprocals of their bootstrap variances. The factors considered were the location–season combination (Berchevaise Island 1991–92, Berchevaise Island 1992–93, ..., Edmonson Point 1995–96), the two stages (guard and crèche), and the interaction between these two factors. Only the location–season factor was significant, due to the large female–male difference at Berchevaise Island in 1994–95. The analysis of variance model was a good fit to the data.

The mean durations were also subjected to an analysis of variance with GLIM, using reciprocals of their bootstrap variances as weights. The factors considered were the location–season combinations, the stage, the sex of the penguins, and the two- and three-factor interactions. The analysis of variance model was a good fit to the data and a highly significant variation associated with the location–season combinations, the stage, the sex of the penguins, and the interaction between the location–season combinations and the stage ($p < 0.001$) was found. The interaction between location–season combinations and sex was also significant ($p < 0.05$), but the interaction between sex and stage and the three-way interaction were not significant ($p > 0.05$). Thus, it was concluded that foraging duration varies with each of the three factors, with the difference between stages and the difference between males and females varying with the site or season. However, the difference between males and females appears to be fairly constant for the guard and crèche stages.

8.4 Analysis of fish acclimation to metals data

8.4.1 Statistical analysis

For Marr et al.'s.[11] challenge 1 experiment, 30 hatchery brown trout (HBT) were randomly selected from 60 of the fish and kept in clean water for 3 weeks. The remaining 30 fish were kept in a solution of metals, with 20% of the concentrations used for the challenge. In the same way, 30 hatchery rainbow trout (HRT) were selected to be kept in the clean water and 29 HRT to be kept in the 20% concentration, while 30 Clark Fork River brown trout (CBT) were selected to be kept in the clean water and 32 CBT to be kept in the 20% concentration. All fish survived the 3-week period.

Following acclimation, the fish were transferred to the challenge solution, which contained nominal solutions of 230 µg/l Zn, 120 µg/l Cu, 3.2 µg/l Pb and 2.0 µg/l Cd. This solution represents concentrations that occur during pulse events in the Clark Fork River, while the 20% solution represents base flow conditions in the river. The survival times of all fish in the challenge solution were recorded.

Marr et al.[11] analyzed their data by a two-factor analysis of variance, with the factors being the three types of fish (HBT, HRT and CBT) and the treatment (control or the 20% acclimation solution). They found a highly significant interaction (p < 0.0001) and therefore interpreted main effects through a series of t-tests with Bonferroni adjusted significance levels to allow for multiple testing. They concluded that the mean time to death for controls was significantly longer for CBT and HRT than for HBT, and CBT also had a significantly higher mean than HBT or HRT when acclimated to the metals. In addition, the mean survival time was significantly higher for the acclimated fish than for the controls for HBT, HRT and CBT. The survival time was not found to be significantly related to the weight covariate.

As an alternative to this conventional type of analysis, a randomization method can be used. For computational convenience, this can be done by setting the analysis up as a multiple regression so that the RT software[14] can be used. The model used can then be

$$Y = \beta_0 + \beta_1 S_2 + \beta_2 S_3 + \beta_4 W + \beta_5 T + \beta_6 S_2 T + \beta_7 S_3 T + \beta_8 S_2 W + \beta_9 S_3 W + e, \quad 8.1$$

where Y is the variable of interest, S_2 is 1 for HRT or otherwise 0, S_3 is 1 for CBT or otherwise 0, W is the weight of the fish (g), T is 0 for controls and 1 for acclimated fish, and e is a random error term that has the same independent distribution for all fish, with a mean of 0. This model allows the expected value of Y to vary with the species, the treatment, and the weight of fish. As well, the effect of the treatment and the effect of weight can vary with the species.

The RT software estimates the regression coefficients β_0 to β_9 using the usual multiple regression equations. It allows two types of randomization. With the first, the significance of the estimated regression coefficients is determined by comparing them with the distributions obtained when the Y values are randomly reallocated to the fish. With the second type, ter Braak's[15] randomization of regression residuals is used. For most sets of data, these two methods give almost the same results. However, a recent extensive simulation study carried out by Anderson and Legendre[16] has demonstrated that, under certain circumstances, ter Braak's method has superior performance. Therefore, this method was used with the fish survival data.

When the Y value used is the survival time of the fish, an examination of regression residuals shows clearly that the error variance tends to increase with the expected value of Y. However, this problem is overcome by using the natural logarithm of survival time as the dependent variable instead of the survival time itself. The randomization analysis was therefore conducted on the logarithms, with 4999 sets of randomized data generated for comparison with the real set. The fitted multiple regression equation was found to be

$$\log_e(X) = 3.08 + 0.0056S_2 + 0.21S_3 - 0.011W + 0.59T - 0.29S_2T$$
$$+ 0.18 S_3T + 0.011S_2W + 0.0022S_3W. \qquad 8.2$$

The coefficient of T is significantly different from zero at the 0.02% level, giving very clear evidence that the acclimation treatment increased the log(survival time). The coefficient of S_2T was significant at the 4.3% level, giving some evidence that the log(survival time) tended to be less for HRT than for HBT. None of the other coefficients are significantly different from zero at the 5% level. As often happens with multiple regression, the conclusions from this randomization analysis are exactly the same as those obtained using the usual t-tests on the regression coefficients.

As an alternative to the multiple regression analysis, a "tailor-made" randomization analysis can be considered. This has also been applied to logarithms of survival times rather than to the actual survival times, because it appears that acclimation increased the variance in survival times, but taking logarithms largely removes the variance differences. The second type of randomization analysis does not require the distribution of survival times to be the same for all three types of fish, and does not require that the relationship between survival times and weight is linear, if any relationship exists. For an analogous test on multivariate data see Clarke.[17]

Consider first the question of whether the acclimation treatment changes the survival times of the fish within each of the three groups HBT, HRT and CBT. This is addressed by the following procedure:

1. The HBT fish are divided up into subgroups on the basis of their weights. This is also done for the HRT and CBT fish.
2. The usual t-statistic is calculated to compare the mean observed survival times for the acclimated and control HBT fish, with a pooled estimate of the variance within the two levels of the treatment. Denote this observed test statistic as t_{11}. Similar t-statistics t_{21} and t_{31} are calculated to compare the observed mean survival times for control and acclimated HRT and CBT fish, respectively. An overall statistic to measure the treatment effect is then $s_1 = \log_e(t_{11}^2) + \log_e(t_{21}^2) + \log_e(t_{31}^2)$.
3. The observed survival times for HBT are randomly reassigned to the fish in the control and acclimated groups within each of the weight subgroups, and the t-statistic is recalculated to give t_{12}. Similar randomized t-statistics t_{22} and t_{32} are calculated for HRT and CBT, together with $s_2 = \log_e(t_{12}^2) + \log_e(t_{22}^2) + \log_e(t_{32}^2)$.
4. Step 3 is repeated a large number N-1 of times to generate N-1 randomized values for each of the test statistics.
5. The observed value of t_{11}^2 is declared to be significantly large at the 5% level if it is larger than 95% of the set of values $t_{i1}^2, t_{i2}^2, ..., t_{iN}^2$, which is evidence that the treatment has an effect for the ith type of fish. Similarly, s_1 is declared to be significantly large at the 5% level if it is larger than 95% of the set of values $s_1, s_2, ..., s_N$, which gives

overall evidence that the treatment affects survival times. Other significance levels are defined in a similar way.

The technique used here of combining individual test statistics to produce an overall statistic for a randomization test appears to have been first suggested by Chung and Fraser[18] for constructing a multivariate test statistic from a series of univariate statistics. It has an element of arbitrariness, but seems reasonable from a common sense point of view.

An important property of the proposed procedure is that it does not require the three types of fish to have the same amount of variation in survival times. Also, only randomizing fish survival times between the control and acclimation treatments within weight subgroups controls for the effects of weight. However, the t-statistics do not explicitly allow for any effects of weight. It is possible that using some alternative form of test statistic based on pairing treated and control fish in the same weight classes would provide a more powerful test. The approach used here has the merit of being simple, and the test is valid because the randomizations that are used do allow for weight to have an effect.

The randomization test is "exact" in the sense that, in the absence of treatment effects, the probability of obtaining a result that is significant at the $100\alpha\%$ level is α. However, for testing for an interaction (the treatment effect varies for the three types of fish), only an approximate randomization test is possible. This approximate test is based on the assumption that, if there were no interaction, the effect of the acclimation treatment is to change the survival time of a fish by a fixed amount D. This overall effect can be estimated by the difference between the mean survival time of all acclimated fish and the mean survival time of all control fish. Then, subtracting the estimated value of D from the survival times of all acclimated fish should produce a modified set of data for which there are no treatment effects for any of the three types of fish. On the other hand, if an interaction is present, subtracting the estimated value of D from the survival times of the acclimated fish will still leave treatment effects for one or more of the types of fish. The idea, then, is to adjust the data using the estimated value of D and then run randomization tests for treatment effects.

The proposed procedure has the following steps:

1. The HBT fish are divided up into subgroups on the basis of their weights. This is also done for the HRT and CBT fish. A constant treatment effect is estimated by the mean survival time of all acclimated fish minus the mean survival time of all control fish. This estimated treatment effect is subtracted from the observed survival time of all acclimated fish.

2. The usual t-statistic is calculated on the modified data to compare the mean observed survival times for the acclimated and control HBT fish, with a pooled estimate of the variance within the two levels of the treatment. Denote this observed test statistic as T_{11}.

Similar t-statistics T_{21} and T_{31} are calculated to compare the observed mean survival times for control and acclimated HRT and CBT fish, respectively. An overall statistic to measure treatment effects is then $S_1 = \log_e(T_{11}{}^2) + \log_e(T_{21}{}^2) + \log_e(T_{31}{}^2)$.

3. The observed survival times for HBT are randomly reassigned to the fish in the control and acclimated groups within each of the weight subgroups. A similar randomization is also carried out for HRT and CBT. A constant treatment effect is then estimated for the randomized data as the difference between the mean survival time for acclimated fish minus the mean survival time for control fish. This estimated treatment effect is then subtracted from the survival time of all the acclimated fish in the randomized data and the statistics T_{12}, T_{22}, T_{32} and $S_2 = \log_e(T_{12}{}^2) + \log_e(T_{22}{}^2) + \log_e(T_{32}{}^2)$ are calculated.

4. Step 3 is repeated a large number N-1 of times to generate N-1 randomized values for each of the test statistics.

5. The observed value of $T_{i1}{}^2$ is declared to be significantly large at the 5% level if it is larger than 95% of the set of values $T_{i1}{}^2$, $T_{i2}{}^2$, ..., $T_{iN}{}^2$, which is evidence that the treatment has an effect for the ith type of fish even after removing the overall estimated effect. Similarly, S_1 is declared to be significantly large at the 5% level if it is larger than 95% of the set of values S_1, S_2, .., S_N, which gives overall evidence for an interaction. Other significance levels are defined in a similar way.

The reason for removing an estimated treatment effect at step 3 is to ensure that the analysis of the randomized data is the same as the analysis of the real data. Note that the analysis would be exact if the estimated overall treatment effects were correct, where these would be 0 for randomized data.

As is the case for the test for any treatment effects, it is possible that some modifications to the details of the proposed test for interaction will provide a more powerful test. For example, in some types of fish, the treatment effect might be close to the average treatment effect for all types of fish, so that T_{ij} is close to 0. This suggests that a combined statistic of the form $Max(T_{1j}{}^2, T_{2j}{}^2, T_{3j}{}^2)$ may be more effective than the sum of logarithms of squares. However, this has not been investigated.

8.4.2 Results

The weight classes used for the randomization procedures just described were 0–20.0g, 20.1–40.0g, 40.1–60.0g, 60.1–80.0g and 80.1–100.0g. The means and standard deviations for the natural logarithms of survival times for control and acclimated fish are shown in Table 8.2.

The test for treatment effects gives the observed statistics $t_{11}{}^2 = 39.21$, $t_{21}{}^2 = 29.12$, and $t_{31}{}^2 = 43.73$ for HBT, HRT and CBT, respectively, leading to an overall statistic of $s_1 = \log_e(39.21) + \log_e(29.12) + \log_e(43.73) = 10.82$. From 4999 randomizations, the t-statistics and s_1 are all significant at the

Table 8.2 Means and Standard Deviations for Logarithms
of Times of Death from Marr et al.'s. Challenge 1 Experiment[11]

Fish	Treatment	n	Mean	SD
HBT	Control	30	3.04	0.30
	Acclimated	30	3.64	0.44
HRT	Control	30	3.32	0.20
	Acclimated	29	3.64	0.25
CBT	Control	30	3.33	0.29
	Acclimated	32	4.09	0.56

0.02% level, showing that, in each case, the value for the observed data was more extreme than what was obtained from any of the randomized sets of data.

For the test for interaction, the observed t-statistics squared are $T_{11}^2 = 0.15$, $T_{21}^2 = 18.44$, and $T_{31}^2 = 2.74$ for HBT, HRT and CBT, respectively, leading to an overall statistic of $S_1 = \log_e(0.15) + \log_e(18.44) + \log_e(2.74) = 2.03$. From 4999 randomizations, the significance level for S_1 is 2.6%, giving some evidence that an interaction was present. For the t-statistics, the significance levels are 60.6% for HBT, 0.02% for HRT, and 2.5% for CBT. It therefore seems that the treatment effect for HBT was similar to the average effect for all fish, the effect for HRT was very different from the average effect for all fish, and the effect for CBT was somewhat different from the average for all fish.

Because the randomization tests just applied to the fish survival data are not standard and the interaction tests are approximate, a small simulation study was run to check whether these tests have the correct type 1 error (probability of a significant result) when the null hypothesis is true, and whether they have reasonable power to detect effects when they exist. This study was based on the available data, as follows:

1. For the HBT fish, the available values for log(survival time) were randomly sampled with replacement to produce a new experimental value for each fish. These values were left unchanged for the control fish, but a treatment effect of Δ_1 was added for each acclimated fish. The same process was carried out for HRT fish, with a treatment of Δ_2 added for acclimated fish, and for CBT fish with a treatment effect of Δ_3 added for acclimated fish. This produced a new data set with distributions for log(survival time) similar to that for the real data and known treatment effects.
2. The randomization tests for treatment effects and interaction were run on the generated data with 249 randomizations, and whether the tests were significant at the 5% level was determined.
3. Steps 1 and 2 were repeated 1000 times for each of several levels for the treatment effects.

Table 8.3 Results from a Simulation Experiment on Randomization Tests for Treatment Effects and Interaction with Data Similar to that from Marr et al.'s Challenge 1 Acclimation Experiment[II]

Treatment effects			% Significant results for tests on treatment effects				% Significant results for tests on interaction			
Δ_1	Δ_2	Δ_3	All	HBT	HRT	CBT	All	HBT	HRT	CBT
			No Effects							
0.00	0.00	0.00	3.8	5.1	5.2	3.8	4.6	4.9	4.1	4.8
			Treatment Effects But No Interaction							
0.15	0.15	0.15	46.1	21.0	50.0	14.8	3.9	4.5	3.7	3.7
0.30	0.30	0.30	94.8	62.7	98.7	46.8	5.7	6.1	5.3	5.2
0.45	0.45	0.45	99.9	93.2	100.0	81.9	4.6	4.8	6.0	4.3
			Interaction With No Treatment Effect for HBT and an Average Effect for HRT							
0.00	0.15	0.30	46.9	4.0	53.5	50.6	24.0	33.3	3.7	26.5
0.00	0.30	0.60	86.0	4.7	98.5	97.3	55.3	81.6	4.8	69.9
0.00	0.45	0.90	95.0	4.2	100.0	100.0	77.7	99.2	4.0	97.1
0.00	0.60	1.20	96.7	5.1	100.0	100.0	84.4	100.0	4.9	99.8
0.00	0.75	1.50	97.8	4.6	100.0	100.0	89.3	100.0	5.0	100.0

The effects of acclimation are to increase the $\log_e$(survival time) by Δ_1 for HBT, by Δ_2 for HRT, and by Δ_3 for CBT. The body of the table shows the percentage of results significant at the 5% level from 1000 simulated sets of data. Cases where the null hypothesis was true are shown in bold type. None of these is significantly different from the desired 5%, at the 5% level of significance (outside the range 3.6%–6.4%). To assess the treatment effects, it can be noted that the standard deviation of the $\log_e$(survival time) for all of Marr et al.'s data is approximately 0.45.

The results from these simulations for a few scenarios are shown in Table 8.3 in terms of the percentage of significant results for tests at the 5% level. Basically, the table shows that, when a null hypothesis was true, the number of significant results was always close to the desired 5%. However, when the null hypothesis was not true, this was often detected with reasonable power. Therefore, the tests have worked well for the simulated situations, although it is an open question as to how the power of these tests compares with the power of other tests of the same hypotheses, including some modifications to the test statistics used for the randomization tests.

8.5   Analysis of goose brood survival data

8.5.1   Statistical analysis

In any 1 year, the data available for estimating the survival function for emperor geese in Alaska consist of records for a number of broods, where each record gives the number in the brood after hatching (day 1), or at a known later age, and the numbers still alive on from one to about 15 later occasions. It is assumed that the expected number of goslings in a brood at t days after hatching is given by a Weibull function of the form

$$E_t = N\exp\{-\alpha t^\Delta\} \qquad\qquad 8.3$$

where N is the number in the brood on day 1. The parameters α and Δ can then be estimated by minimizing the sum of squared differences between the observed numbers in broods and the numbers expected from this equation.

To assess the accuracy obtained by this rather *ad hoc* method of estimation, 1000 bootstrap sets of data were generated for which the number of broods was kept equal to the number for the real data, but with these broods chosen by randomly resampling the observed broods with replacement. This method of bootstrapping accounts for the possibility that survival probabilities vary from brood to brood, and has been proposed before by Flint et al.[19] for estimating a survival function using the Kaplan and Meier[20] and Mayfield[21,22] methods.

To compare the survival in 2 or more years, one possible approach involves fitting both a separate survival function for each of the years and also a common survival function for all of the years. The reduction in the error sum of squares due to fitting the function separately for each year can then be calculated to give the value S_1. A randomized set of data can then be calculated by randomly reallocating the brood records to years, keeping the number of broods equal to that for the observed data, and calculating S_2, the reduction in the error sum of squares due to fitting the survival function separately for each year. Repeating the randomization M-1 times gives the sequence of statistics $S_1, S_2, ..., S_M$. The hypothesis that the survival function is the same in each year is then tested at the $100\alpha\%$ level by seeing whether S_1 exceeds $100(1-\alpha)\%$ of the complete set of M statistics.

This procedure is similar to one used to compare the growth of Pigeon Guillemot chicks on two islands in Prince William Sound following the *Exxon Valdez* oil spill in 1989 (Example 13.2 in Reference 10). Therefore, it will not be considered further here.

8.5.2 Results

For 1993, results are available on the survival for goslings in 83 broods and these give the survival function estimated by non-linear least squares to be

$$E_t = \exp(-0.133t^{0.470}) \qquad\qquad 8.4$$

so that the estimates of α and Δ are 0.133 and 0.470, respectively. When 1000 bootstrap samples were generated and the survival function reestimated, it was found that the mean and standard deviation for the estimates of α were 0.157 and 0.103, respectively, which therefore suggests that the estimator of α is biased upward by about 0.024 (the difference between the bootstrap "true" value of 0.133 and the mean of the estimates). The mean and standard deviation for the bootstrap estimates of Δ were 0.475 and 0.217, respectively, suggesting a bias in estimation for this parameter of about 0.004. The correlation between the estimates of α and Δ from the bootstrap samples was substantial at -0.91.

Basically, this analysis suggests that 83 broods is not enough to ensure accurate estimation of the survival function, as is demonstrated clearly by the top part of Figure 8.1, which shows a plot of the estimated survival function from 25 bootstrap samples. The variation between these functions is considerable.

In 1994, results were available for 72 broods. The estimated survival function is

$$E_t = \exp(-0.593t^{0.176}) \qquad\qquad 8.5$$

and 1000 bootstrap samples produced the following results: the mean of the estimates of α is 0.612, with a standard deviation of 0.149; the mean of the estimates of Δ is 0.175, with a standard deviation of 0.068; and the correlation between the estimates of α and Δ is -0.81. Thus, the estimation of α seems to have a bias of approximately 0.019, and the estimation of Δ almost no bias. Plots of 25 survival functions estimated from bootstrap samples are shown in the bottom part of Figure 8.1. There is less variation than there is for 1993.

With a bootstrap analysis of this type, a number of questions related to the extent to which the bootstrap samples mimic real resampling of the population remain unanswered. However, at least as a first approximation,

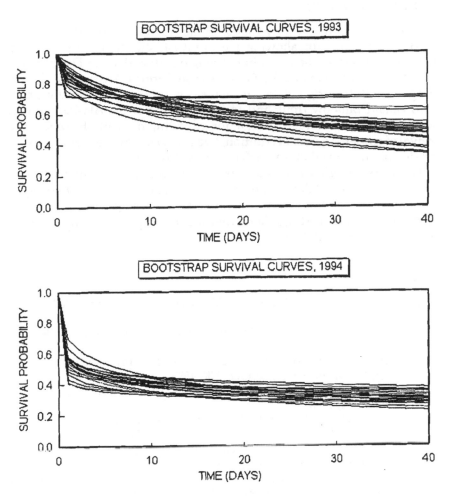

Figure 8.1 Bootstrap estimates of survival curves for goslings in 1993 and 1994.

they give a good indication of the amount of variation that can be expected from sampling errors.

8.6 Conclusions

The three examples that have been discussed demonstrate the way in which the computer-intensive methods of randomization and bootstrapping can be used to develop special purpose analyses for ecological time to event and survival rate data that exhibit a complicated correlation structure, heterogeneity of variances, and non-standard distributions. More-general information about the use of randomization and bootstrap methods more generally can be found in a number of recent books on these topics.[14,23-25]

Acknowledgements

I am most grateful to Dr. Judy Clarke of the Australian Antarctic Division, Kingston, Tasmania, for allowing me to investigate the use of computer-intensive methods on the penguin foraging duration data; to Dr. John Marr of the Mississippi-Alabama Sea Grant, Ocean Springs, Mississippi, for providing me with the original data from the experiment on the survival times of brown and rainbow trout; and to Joel Schmutz of the Alaska Biological Sciences Center, Anchorage, Alaska, for providing me with the data on the survival of emperor geese. In addition, two referees provided comments that considerably improved an earlier version of this chapter.

The work reported here was partially funded under contract CO1616 from the New Zealand Foundation for Research, Science and Technology for collaborative research between the University of Otago and the National Institute of Water and Atmospheric Research on Improved Statistical Methods for Environmental Assessment.

References

1. Cox, D.R. and Oakes, D., *Analysis of Survival Data*, Chapman & Hall, London, UK, 1984.
2. Lee, E.T., *Statistical Methods for Survival Data Analysis*, 2nd ed., Wiley, New York, NY, 1992.
3. Burgman, M.A., Ferson, S. and Akçakaya, H.R., *Risk Assessment in Conservation Biology*, Chapman & Hall, London, UK, 1993.
4. Ferson, S. and Akçakaya, H.R., *RAMAS/Age User Manual*, Exeter Software, Setauket, NY, 1990.
5. Akçakaya, H.R. and Ferson, S., *RAMAS/Space User Manual: Spatially Structured Population Models for Conservation Biology*, Exeter Software, Setauket, NY, 1990.
6. Manly, B.F.J., *Stage-Structured Populations: Sampling, Analysis and Simulation*, Chapman & Hall, London, UK, 1990.
7. Pollock, K.H., Nichols, J.D., Brownie, C. and Hines, J.E., Statistical inference for capture–recapture experiments, *Wildlife Mono.*, 107, 1990.
8. Manly, B.F.J., *Popsys P1f, Stage-Frequency Analysis*, Ecological Systems Analysis, Pullman, WA, 1990.
9. Smith, S.G., Skalski, J.R., Schlechte, J.W., Hoffman, A. and Cassen, V., *SURPH1 Manual: Statistical Survival Analysis of Fish and Wildlife Tagging Studies*. Center for Quantitative Science, School of Fisheries, University of Washington, Seattle, WA, 1994.
10. Manly, B.F.J., *Randomization, Bootstrap and Monte Carlo Methods in Biology*, 2nd ed., Chapman & Hall, London, UK, 1997.
11. Marr, J.C.A., Bergman, H.L., Lipton, J. and Hogstrand, C., Differences in relative sensitivity of naive and metals-acclimated brown and rainbow trout exposed to metals representative of the Clark Fork River, Montana, *Can. J. Fish. & Aquat. Sci.*, 52, 2016, 1995.
12. Clarke, J., Manly, B.F.J., Kerry, K., Franchi, E. and Focardi, S., Sex differences in Adélie penguin foraging strategies, *Polar Biol.*, 20, 248, 1998.

13. Francis, B., Green, M. and Payne, C., *The GLIM System, Release 4 Manual*, Clarendon Press, Oxford, UK, 1993.
14. Manly, B.F.J., *RT, A Program for Randomization Testing, Version 2.1*, Centre for Applications of Statistics and Mathematics, University of Otago, Dunedin, New Zealand, 1997.
15. ter Braak, C.J.F., Permutation versus bootstrap significance tests in multiple regression and ANOVA, in *Bootstrapping and Related Techniques*, Jöckel, K.H., Ed., Springer-Verlag, Berlin, Germany, 1992, 79.
16. Anderson, M.J. and Legendre, P., An empirical comparison of permutation methods for tests of partial regression coefficients in a linear model, *J. Stat.Comp. & Simulation*, 62, 271, 1999.
17. Clarke, K.R., Non-parametric multivariate analyses of changes in community structure, *Aust. J. Ecol.*, 18, 117, 1993.
18. Chung, J.H. and Fraser, D.A.S., Randomization tests for a multivariate two-sample problem, *J. Am. Stat. Assoc.*, 53, 729, 1958.
19. Flint, P.L., Pollock, K.H., Thomas, D. and Sedinger, J.S., Estimating prefledgling survival: allowing for brood mixing and dependence among brood mates, *J. Wildlife Mgmnt.*, 59, 448, 1995.
20. Kaplan, E.L., Meier, P., Nonparametric estimation from incomplete observations, *J. Am. Stat. Assoc.*, 53, 457, 1958.
21. Mayfield, H., Nesting success calculated from exposure, *Wilson Bull.*, 73, 255, 1961.
22. Mayfield, H., Suggestions for calculating nesting success, *Wilson Bull.*, 87, 456, 1975.
23. Edgington, E.S., *Randomization Tests*, 3rd ed., Marcel Dekker, New York, NY, 1995.
24. Efron, B. and Tibshirani, R.J., *An Introduction to the Bootstrap*, Chapman & Hall, New York, NY, 1993.
25. Westfall, P.H. and Young, S.S., *Resampling-Based Multiple Testing*, Wiley, New York, NY, 1993.

chapter 9

Time to event analysis in engineering

Alan Kimber

Contents

9.1 Introduction

This chapter presents a brief, non-technical review of time to event methods in engineering, particularly reliability engineering. Mathematical formulae have been avoided throughout. An outline of the basic parametric modeling approach is set out together with some topics of current research interest, including random effects modeling, technological evolution and degradation modeling. Where possible, potential links between time to event methods in engineering and environmental toxicology are highlighted.

9.2   Use of time to event methods in engineering

Time to event (TTE) methods have been used extensively and successfully in engineering for many years, particularly in the context of reliability of components and systems where the event of interest is usually failure of the component or system to perform its designated task. Such failures may be catastrophic (e.g., failure of an aero-engine part causes an explosion of the engine) or nominal (e.g., failure of the same engine part might be defined in terms of engine efficiency loss: once the efficiency falls below a certain predetermined level the part is deemed to have failed, even though the engine is still functioning). Clearly, in such a context, short times to failure are of concern. However, the event of interest need not always be bad. TTE methods are used to analyze repair times, in which case, very long times to complete repairs are of concern. Furthermore, in areas such as materials engineering, the "times" to events need not be times at all but some other inherently positive quantities, such as the strength of a specimen of material.

In these applications, a practical issue is that the observation period in any experiment or study is necessarily finite. Thus, in practice, it is often the case that not all components, or other materials of interest, will have failed during the observation period, thereby giving rise to right censored observations. That is, if one observes a component for t time units at which point the component has not failed, then it is known only that its time to failure exceeds t units. This problem of right censoring is a common feature of all practical TTE analysis, whether the application is in engineering, medical studies or environmental science.

The link between TTE methods in engineering and applications in environmental toxicology is clear, with components or systems replaced by living things, catastrophic failure replaced, for example, by death, and nominal failure, for example, replaced by some other measure of failure to thrive. Of course, the event of interest in a TTE analysis can be a more positive outcome, such as completion of a repair in the engineering context or recovery in a medical or biological set-up. Thus, it is to be hoped that the discussion of TTE methods in engineering offered in this chapter will be relevant to environmental toxicologists.

The intention is to present a non-technical discussion of TTE methods in engineering with a bare minimum of mathematical formulae. Appropriate references will be given to enable those who would like to know the technicalities and mathematical detail to find out more. The basic methods of TTE analysis in engineering will be discussed, but some current research topics (that the author hopes may have some bearing on environmental toxicology) will also be featured. It is assumed that fundamental quantities in TTE analysis, such as the hazard function, are familiar to the reader (they are covered elsewhere in this volume). References in which the fundamentals are covered in the reliability context are Crowder et al.[1] and Ansell and Phillips.[2]

9.3   The basics

TTE methods have been used in engineering for many years. An early example is Davis,[3] who analyzed times to failure of bus motors. In the early days, the lifetime of a component was often assumed to be exponentially distributed (thereby modeling the occurrence of failures as a Poisson process) but, not surprisingly, modeling a possibly complicated failure process by a one-parameter distribution was often, at best, a rather crude approximation of reality. For example, this approach does not allow for the possibility of "wear-out" (or, in actuarial terms, an increasing force of mortality). Therefore, other more flexible distributions were used increasingly as the basic building blocks in TTE analyses. Typically, these are still simple models, usually involving two parameters but with some kind of theoretical justification in their favor. A major example is the two-parameter Weibull distribution.[4] This includes the exponential distribution as a special case and has an extreme value justification. For example, consider a chain that is to bear a load. The chain will tend to break at its weakest link. Thus, one can regard the chain as a large number of links and its strength is essentially determined by the lowest strength of all the links. Results from probability theory indicate that the distribution of the minimum of a set of quantities (strengths, for example) has a particular limiting form (just as the sum of a large number of quantities has a limiting normal form). The Weibull distribution satisfies this limiting form for minima. Further details can be found in, for example, Crowder et al.[1]

The inverse Gaussian distribution also has an engineering justification. This, too, is a two-parameter distribution, but one that does not include the exponential as a special case. The motivation for this is as follows: if the performance of a component degrades over time and the component fails when the degradation reaches a given level, the failure time is the time at which the degradation process first crosses this level (a so-called first passage time distribution). If the degradation process is of a particular type (a Wiener process), the time to failure has an inverse Gaussian distribution. A useful reference to this distribution is Johnson and Kotz,[5] who give mathematical details, together with some simple statistical methods for inference.

Other popular two-parameter distributions that are sometimes used include the lognormal distribution, the gamma distribution and the log-logistic distribution. The engineering justification for these is perhaps weaker than for the Weibull and inverse Gaussian distributions. Good discussions of the lognormal and gamma distributions are to be found in Johnson and Kotz[5] and Crowder et al.[1] Cox and Oakes[6] discuss the log-logistic distribution.

Note that the natural starting point in TTE analysis in engineering is parametric modeling. This is in contrast, for example, to TTE analysis in medical studies, where the starting point is often nonparametric or semi-parametric. An explanation for this is that the underlying failure process of a ball bearing is likely to be less complex than the onset of organ rejection in a heart transplant patient, thereby making the reliability analysis more amenable to simple parametric description. Moreover, in principle at least, there may be some well understood physics underlying an engineering process (crack growth, for example) that can be incorporated into a parametric model, whereas such opportunities may be less clear cut in other spheres such as medicine. Further, in reliability engineering, aspects of the failure time distribution *per se* are of major importance in themselves so that good models for the failure time distribution are of particular value. However, in medical statistics, key questions often relate to comparison of treatment effects so that fine detail of the time to event distribution is of secondary importance. Hence, there is no need to build a possibly complex model for failure times in that context. In environmental toxicology, given the range of applications involved, one might anticipate that both approaches would have a place.

In engineering, typical quantities of interest are: the reliability value (that is, the probability that a component survives for longer than a specified time); a specified quantile of the lifetime distribution (for example, the median lifetime); and, rather similar to many environmental toxicology application, the value taken by some stress condition that leads to some given, fixed proportion of failures by a specified time.

Simple parametric models can be generalized to include covariate information (such as stress levels, temperature and design information), thereby yielding parametric regression models. Two-parameter distributions that are location-scale models on some scale are particularly convenient, such as the Weibull and lognormal distributions, both of which are location-scale models on the log-scale. By analogy with standard (Normal-theory) regression models, a natural starting point is to include covariate information that affects the location but not the scale (rather like the constant variance assumption in classical analysis of variance). Of course, more complex approaches are possible but these relatively straightforward regression models are a good place to begin.

So, to sum up this section, in engineering the natural starting point for TTE analysis is a relatively simple lifetime distribution, possibly augmented by covariate information. Checks on model adequacy are important and frequently involve simple graphical methods such as probability plots, and non-parametric methods such as the Kaplan-Meier or product limit estimator; see, for example, Crowder et al.[1] or Ansell and Phillips[2] for a discussion of these methods.

9.4 Censoring

As mentioned above, an almost inevitable aspect of TTE analysis is the presence of incomplete, or censored, observations. Observations may be right-, left-, or interval-censored, therefore it is useful to discuss these different cases here.

For a right-censored observation it is known only that the actual failure time is greater than some observed value. This might arise because observation takes place for only a finite time, so that some components may not have failed by the end of the observation period. This is the most common type of censoring that is acknowledged in TTE engineering work.

A less common problem in engineering is the presence of left-censored observations, for which it is known only that the lifetime is less than some observed value. This might arise if the component failed between the start of the experiment and the first time the component's status was checked.

An observation is interval censored if its time to failure is known to lie only within some given interval. Right censoring and left censoring can be regarded as special cases of interval censoring. For example, if the upper end point of the interval is infinity, then an interval-censored observation is, in fact, just right censored. Further, strictly speaking, all lifetimes are interval censored because all measurements are given to finite accuracy. However, this is of little practical importance unless this lack of accuracy is severe, though particular care is needed for observations near zero. For example, accuracy to one decimal place might be sufficient for longer lifetimes but inadequate at low lifetimes, particularly if a log-linear analysis is to be performed.

A particularly coarse form of censoring arises in so-called current status data. Here, items are put on test at time zero and subsequently observed only after a fixed time, at which each item has either failed or is unfailed. Thus, it is known only whether each item is "alive" or "dead" at the specified time. Hence, the times to failure of failed items are left censored and the times to failure of unfailed items are right censored. This situation also arises in frequently used ecotoxicological tests, for example, 96-h tests on fish. Note that the difference between current status data and the (in engineering) much more common "usual" TTE framework is that, in a standard TTE analysis, full information on lifetimes of failed items is available, whereas only the coarser and less informative left-censored values are available in current status data.

Data with only right-censored or failed items may be handled statistically quite simply. Statistical methods for data with more complicated patterns of censoring are also well developed. Some relevant methods for handling censored data are given in the next section. From the point of view of the practitioner, the availability of powerful packages such as S-Plus means that TTE models can be fitted to censored data quite routinely.

9.5 Statistical methods

In engineering, the computation required to fit statistical models such as a Weibull regression model has gone from being time consuming to being trivial (most standard statistical packages will do this). Consequently, the statistical methods used have evolved from "quick and dirty" *ad hoc* methods (simple least squares, for example) to methods that make full and efficient use of the available data. Typically, methods based on the likelihood are used to obtain maximum likelihood estimates of quantities of interest. To obtain standard errors or confidence intervals for these quantities or to perform hypothesis tests, standard large sample methods based on the information matrix (see, for example, Crowder et al.[1]) are often used, though they may be poor in small samples or heavily censored samples. Alternatively, more computer-intensive methods, such as the bootstrap (see, for example, Efron and Tibshirani[7]) are used. Bayesian methods are also used on occasion, although it is fair to say that the Bayesian approach has yet to be used routinely in TTE analysis in engineering. A discussion of the basics of Bayesian methods in the present context is given in Crowder et al.[1]

To check the goodness-of-fit of a chosen statistical model and to investigate the appropriateness of underlying assumptions, simple graphical and numerical methods are used. In the case of data in which there is no censoring other than right censoring, diagnostic methods are well developed and straightforward.[1,2] Examples include the Kaplan-Meier estimator and probability plots. Current status data is also easy to handle because binary data methods (logistic regression, for example) can be used efficiently to fit appropriate models to the data. Note that Weibull, log-logistic and lognormal models in the TTE framework are clearly related to complementary log-log, logistic and probit binary regression models. A good discussion of diagnostics in this case is given in Collett.[8] For data in which the censoring is more complex than the two special cases of right censoring only and current status data, there are some useful results for diagnostics in Turnbull[9] and, more recently, in Pan and Chappell.[10]

In most basic statistics courses it is usually pointed out that, for any statistical analysis to be reasonably powerful, a large enough sample must be taken. This is not the whole story, however, when there is censoring. Even if a large number of items are put on test (hence a large sample size), if the vast majority of observations are censored, then the data will contain relatively little information (in a statistical sense). This, in turn, may mean that important quantities of interest can be only rather imprecisely estimated. The situation is worse if the initial number of items on test is not large. Unfortunately, this is just the type of situation that frequently crops up in engineering. The number of items on test may be relatively few because of expense of manufacture or because there is not the time, space or staff to put large numbers on test. Time may be tight for testing, thereby making a large proportion of right-censored observation highly likely. Furthermore, if the engineering has been successful, one would expect items to be highly reliable, once again making a large proportion of right-censored observations highly likely. Similar remarks might also apply in medical or biological TTE studies where, for example, ethical concerns could constrain the number of individuals who can take part in the study.

A traditional approach to circumventing this kind of problem is to use acceleration factors. That is, items that are put on test are subject to extreme conditions or stresses to ensure that a large proportion of items fail. The good thing here is that inferences can be made reasonably precisely. However, extrapolation is needed to make inferences about low stress parameters.

It is not difficult to see that this can be a risky thing to do. First, errors propagate, so extrapolation may give very imprecise results for the real parameters of interest. Second, in engineering, there is often more than one failure mode and different failure modes may dominate at different stress levels. For example, consider a car tire. Under mild conditions, the main failure mode may be that the tread depth becomes too shallow (a nominal failure). However, in accelerated conditions (high stress) it may be that the tire tends to puncture (a catastrophic failure) long before tread depth becomes a problem. Thus, an accelerated test in this example would give little information about the failure mode of interest, thereby giving potentially misleading reliability information about the tire in mild conditions. Therefore, accelerated testing should be used with caution, ideally when knowledge of the failure process, including the various modes of failure, indicates that appropriate results are possible.

Similar concerns might be an issue in environmental toxicology where high dose effects over a short time period might be extrapolated to estimate low dose effects over a long period. Again, different failure modes might occur here (death at high dose levels or failure to reproduce at low dose levels).

These concerns apply to any situation in which extrapolation is used. However, they highlight the need to collect enough of the right kind of information and to make the best possible use of this information. This is an issue that arises again in sections below.

9.6 Topics of current interest

We now consider some areas of current statistical research that have relevance to TTE analysis and may be useful in engineering applications. These are random effects models for TTE distribution, Bayesian methods in technological evolution, and degradation modeling. These are not the only major relevant research areas but are those with which I have been most involved in recent years. I will touch on other growth areas at the end of this section.

9.6.1 Random effects in TTE data

Suppose, for example, we wish to perform a TTE analysis on lifetimes of ball bearings. To all intents and purposes, these bearings are homogeneous. If they were put on test under identical conditions, we would expect them to behave similarly. Of course, the lifetimes would exhibit natural variability, but this could be modeled in a straightforward way by using, say, a Weibull or lognormal model. In contrast, if we wish to perform a TTE analysis on remission times of cancer patients undergoing a treatment, then, although considerable covariate information might be available (for example, sex, age and size of tumor at diagnosis) we would not be surprised to observe extra variability in such data over and above that which could be explained by a simple Weibull or lognormal model. Essentially, there would be a patient-specific effect. A neat way of dealing with this would be to postulate a distribution for this across patients and treat it as a random effect. Two readable papers dealing with this approach are Hougaard[11] and Crowder.[12]

Note that a Weibull distribution is itself a random effects distribution: an appropriate mixture of exponential distributions has a Weibull distribution with decreasing hazard function, and an appropriate mixture of Weibull distributions itself has a Weibull distribution (see Crowder[13] for more details).

Let us return now to the engineering context. Today, materials and components are produced whose structure and fabrication are orders of magnitude more complex than those of ball bearings (fiber composites and computer chips, for example). Thus, one might anticipate that random effects modeling in a parametric TTE analysis might be of value for such items. Recent work in this area includes Kimber[14] and Crowder and Kimber.[15] If TTE analysis were to be used routinely in environmental toxicology, then there is clearly scope for random effects modeling to cope with natural heterogeneity of populations.

9.6.2   Technological evolution

We have already discussed the problem of data that contain relatively little information. One situation in which this is particularly acute is technological evolution. Suppose that good information on the reliability of a particular item or component is available from previous TTE analyses and, possibly, from other sources. A customer requires a small modification to the design and wants the resulting modified items in a hurry, leaving little time for testing of the new item. This is an example of the technological evolution problem. If the design change is minor (for example, the new item is painted pink instead of orange), knowledge about the original version can be used to augment the very limited TTE analysis of the modified item in a natural way via a Bayesian approach. A particularly simple case in which component lifetimes are exponentially distributed was covered in Whitmore et al.,[16] who used conjugate prior information that utilized engineering judgment to sharpen inferences about the reliability of the modified item. Young[17] adopted a similar approach to this problem, but she used nonconjugate prior information to incorporate engineering judgment.

Work on this type of problem is at an early stage, but does highlight the potential for using subject matter knowledge to augment results from TTE or other experiments. There may be scope for this type of approach in the case of ecotoxicological studies in which information from earlier tests can be used to improve inferences from later tests.

9.6.3   Degradation modeling

In engineering, the performance of a component or system will tend to degrade over time with use. This degradation may or may not be monotonic (that is, the performance never improves), but usually there is a trend toward increasing degradation. Suppose the component or system fails as soon as it crosses a critical level so that the failure time will have a first passage time distribution. If the degradation process can be observed or modeled, there is scope for taking TTE analysis one step further and using this additional information.

Recall that, in current status data, there is only incomplete (that is, censored) information available for both failed and unfailed items. In standard TTE analysis, complete information is available on the failed items, but the only information on unfailed items is incomplete. With a degradation analysis, complete information (the failure time) is available on the failed items, and more detailed information is available on the unfailed items; the degradation status of unfailed items will give an indication of how close to failure the unfailed items were. For example, consider two unfailed items, one of which has seriously degraded and is near the failure level by the end of the experiment; the other has hardly degraded at all. In a standard TTE analysis without the degradation measure, both of these would be recorded simply as unfailed items and would make the same contribution to the analysis even though, in fact, one was close to failure and the other one was not.

Whilst degradation models in engineering have been considered widely in the literature, there has been little in the way of using efficient statistical approaches to fit these models to actual degradation data. Two recent noteworthy exceptions are the papers by Whitmore[18] and Whitmore and Schenkelberg.[19] In the context of ecotoxicology, if organisms are under test in a TTE context and if their degree of aging or some other measure of health can be monitored, then this degradation information could be used to augment the information gained from the TTE aspect of the study.

9.6.4   Other topics

Many other topics are major research areas in TTE analysis in engineering. Some of these will be outlined briefly here with references.

The problem of multiple failure modes was previously mentioned. In situations where typical TTE information is combined with the particular modes that resulted in the failures, a so-called competing risks analysis is called for. The basics of this approach are discussed briefly in Crowder et al.[1] There is considerable active research in this topic, but much of it is rather technical. Within the context of environmental toxicology, there is clearly scope for differentiating between different failure modes so that competing risks analysis is a possibility.

Other areas of interest within engineering include repairable systems reliability including software reliability,[20] general system reliability[21] and load sharing problems.[22] It is perhaps unlikely that these topics have any direct relevance to environmental toxicology.

9.7   Discussion

First, some general remarks about statistics: The subject continues to grow rapidly, and the scope for their being used constructively to help solve complex problems has increased dramatically in recent years. One reason for this is the enormous computing power that is now available, thereby opening the way to routine use of computer-intensive methods such as bootstrapping (Efron and Tibshirani,[7] and Chapter 8 in this volume), and Markov Chain Monte Carlo methods.[23]

TTE methods and related topics are very active research areas that are developing rapidly. Two readable texts that give a sound introduction to TTE methods in engineering are Crowder et al.[1] and Ansell and Phillips.[2] The journals *IEEE Transactions in Reliability* and *Technometrics* contain numerous papers on engineering-based statistics, many of which are concerned directly with TTE methods. The journal *Lifetime Data Analysis* is devoted solely to TTE and related methods (in all disciplines, not just engineering). More biological TTE methods and applications are frequently to be found in *Biometrics* and *Statistics in Medicine*. TTE papers also appear in a variety of other statistical journals, including those of the *Royal Statistical Society* and of the *American Statistical Association*.

Acknowledgments

I should like to thank the SETAC U.K. conference organizers for inviting me to give this paper. My thanks also go to the referees for their helpful and encouraging comments. Part of this work was funded by an EPSRC research grant.

References

1. Crowder, M.J., Kimber, A.C., Smith, R.L. and Sweeting, T.J., *Statistical Analysis of Reliability Data*, Chapman and Hall, London, UK, 1991.
2. Ansell, J.I. and Phillips, M.J., *Practical Methods for Reliability Data Analysis*, Oxford University Press, Oxford, UK, 1994.
3. Davis, D.J., An analysis of some failure data, *J. Am. Stat. Soc.*, 47, 113, 1952.
4. Weibull, W., A statistical distribution function with wide applicability, *J.Appl. Mech.*, 18, 293, 1951.
5. Johnson, N.L. and Kotz, S., *Distributions in Statistics: Continuous Univariate Distributions 1*, Wiley, New York, NY, 1970.
6. Cox, D.R. and Oakes, D., *Analysis of Survival Data*, Chapman and Hall, London, UK, 1984.
7. Efron, B. and Tibshirani, R.J., *An Introduction to the Bootstrap*, Chapman and Hall, London, U.K., 1993.
8. Collett, D., *Modelling Binary Data*, Chapman and Hall, London, U.K., 1991.
9. Turnbull, B.W., The empirical distribution function with arbitrarily grouped, censored and truncated data, *J. Royal Stat. Soc. B*, 38, 290, 1976.

10. Pan, W. and Chappell, R., A nonparametric estimator of survival functions for arbitrarily truncated and censored data, *Lifetime Data Analysis*, 4, 187, 1998.

11. Hougaard, P., Life table methods for heterogeneous populations: distributions describing the heterogeneity, *Biometrika*, 71, 75, 1984.

12. Crowder, M.J., A distributional model for repeated failure time measurements, *J. Royal Stat. Soc. B*, 47, 447, 1985.

13. Crowder, M.J., A multivariate distribution with Weibull connections, *J. Royal Stat. Soc. B*, 51, 93, 1989.

14. Kimber, A.C., A Weibull-based score test for heterogeneity, *Lifetime Data Analysis*, 2, 63, 1996.

15. Crowder, M.J. and Kimber, A.C., A score test for the multivariate Burr and other Weibull mixture distributions, *Scan. J. Stat.*, 24, 419, 1997.

16. Whitmore, G.A., Young, K.D.S. and Kimber, A.C., Two-stage reliability tests with technological evolution: a Bayesian analysis, *Appl. Stat.*, 43, 295, 1994.

17. Young, K.D.S., A Bayesian analysis of updated component data, *The Statistician*, 43, 129, 1994.

18. Whitmore, G.A., Estimating degradation by a Wiener diffusion process subject to measurement error, *Lifetime Data Analysis*, 1, 307, 1995.

19. Whitmore, G.A. and Schenkelberg, F., Modelling accelerated degradation data using Wiener diffusion with a time scale transformation, *Lifetime Data Analysis*, 3, 27, 1997.

20. Ascher, H. and Feingold, H., *Repairable Systems Reliability*, Marcel Dekker, New York, NY, 1984.

21. Natvig, B., Sormo, S., Holen, A.T. and Hogasen, G., Multistate reliability theory – a case study, *Advances in Applied Probability*, 18, 921, 1986.

22. Wolstenholme, L.C. and Smith, R.L., Statistical inference about stress concentrations in fibre-matrix composites, *J. Mat. Sci.*, 24, 1559, 1989.

23. Gilks, W.R., Richardson, S. and Spiegelhalter, D.J., *Markov Chain Monte Carlo in Practice*, Chapman and Hall, New York, NY, 1996.

chapter 10

Can risk assessment be improved with time to event models?

Mark Crane, Peter F. Chapman, Tim Sparks, John Fenlon and Michael C. Newman

Contents

1-56670-582-7/02/$0.00+$1.50
© 2002 by CRC Press LLC

10.1 Introduction

10.1.1 What is time to event?

Most ecotoxicity tests include at least some measurements that occur uniquely in time. An organism can die only once. Hatching, development through irreversible stages, and production of the first batch of offspring are other examples of events that will only occur once in an organism's lifetime. If the time taken for this event to occur is recorded for each organism in a test, a group of statistical techniques, collectively known as time to event analyses, can be used to describe the shape of the response and compare these shapes among different treatments.

10.1.2 Background to the Wellesbourne workshop

The use of certain statistical techniques to analyze the results of ecotoxicity tests has been criticized for several years. Skalski[1] was among the first to point out that the commonly calculated no observed effect concentration (NOEC) is neither a safe nor statistically robust summary of ecotoxicity data. Other authors since then have made similar points.[2-9] However, despite widespread concern over the use of the NOEC, a suitable replacement statistic had not been clearly identified.

The Organization for Economic Cooperation and Development (OECD) is responsible for promoting the international harmonization of testing procedures through the production of ecotoxicity test guidelines.[10] Officials in OECD were aware of the NOEC debate and commissioned a report on possible alternatives.[11] This report recommended a move away from the NOEC and toward the use of dose-response modeling to estimate an effective concentration (EC) value. A workshop held at Royal Holloway, University of London, U.K. in April 1995 supported this recommendation.[12] However, it was also noted that estimation of a time-specific EC value (e.g. a 96-h EC50) ignored toxicity trends through time.

A subsequent workshop organized by OECD in Braunschweig, Germany, in October 1996 reinforced the view that time was an under-described dimension in the analysis of ecotoxicity data.[13] This was considered unfortunate, as an increasing number of researchers had shown that incorporation of time into ecotoxicity analyses could enhance interpretation and improve risk assessment.[14-18]

Despite these calls for the use of temporal observations in ecotoxicity analyses, the use of time to event approaches in other scientific and engineering disciplines,[19] the existence of appropriate computer software,[20] and the publication of authoritative statistical texts on the subject,[21,22] there was still no forum in which the ecotoxicological relevance of time to event approaches had been fully discussed. There was also a lack of suitable publications for non-mathematical ecotoxicologists that explained the methods, advantages and disadvantages of time to event analyses.

A workshop was held April 2–3, 1998 at Horticulture Research International in Wellesbourne, Warwickshire, U.K. to address the issues raised above. Twenty-seven delegates from eight countries participated in the workshop (see Appendix). Delegates with backgrounds in ecotoxicology or statistics came from industry, academia and government regulatory agencies.

Four working groups were set up to discuss the implications of time to event analyses for the predictive assessment of pesticides and industrial chemicals, and the retrospective assessment of industrial effluents and environments historically contaminated through accidental spillages or industrial use of chemicals. After introductory presentations from delegates with an understanding of environmental regulation in North America and Europe, the working groups considered the following questions:

- The Regulatory Framework
  - Are the laws that regulate hazardous substances sensible?
  - Do the regulators of hazardous substances interpret the laws correctly?
- Ecotoxicity Test Methodology
  - Do current test methods provide useful information?
  - Could more useful information be obtained from current test methods?
  - Would alternative test designs provide more useful information?
- Chemical Risk Assessment
  - Is information from ecotoxicity tests used optimally by chemical risk assessors?
- Time to Event Methods
  - How can time to event approaches be used to improve the interpretation of ecotoxicity tests?
  - Should current test designs be modified to optimize time to event analyses?
  - Should new tests be designed to allow time to event analyses?

The aim of these questions was first to establish whether there was a clear legal mandate for environmental protection and, if so, what its objectives were. Having established this foundation, the goal was then to explore the most useful techniques for achieving the established objectives.

10.2 The regulatory framework

10.2.1 Are the laws that regulate hazardous substances sensible?

Environmental laws relating to the regulation of hazardous substances appear to provide a sensible framework, so long as they remain broad, general and aspirational. This is because they are then flexible and allow environmental regulators considerable freedom of interpretation (although whether such freedom is always exercised correctly is an arguable point). Most environmental laws in democracies have evolved through time as a result of pressures from different interest groups. They now tend to allow either the manufacturers or dischargers of hazardous substances and the public or their representatives to appeal against the decisions made by regulators. However, laws that specify the use of poor techniques are clearly of limited use. For example, the European Union Directive on Plant Protection Substances (Directive 91/414/EEC) specifies the use of the NOEC for risk assessment. Legal provisions for data confidentiality can also present a problem when environmental regulators are unable to use data submitted for one purpose (e.g., pesticide registration) for another purpose (e.g., setting an Environmental Quality Standard).

10.2.2 Do the regulators of hazardous substances interpret the laws correctly?

The interpretation of environmental laws by environmental regulators appears generally to be correct, but regulators in different countries may take different views, especially if data on substances such as pesticides are marginal. This seems appropriate in democracies, because these different views may reflect the economic, political and land use interests in different nations. For example, regulators in the U.K. are obliged to consider cost–benefit analyses, while those in some other European countries are not.

    Some difficulties exist in the way regulators interpret environmental laws. First, there is often no consensus between scientists and the public over what environmental values should be protected. This can translate into an inability of regulators to formulate operational risk assessment goals. Such goals are often defined in terms of, "What can we measure?" rather than, "What do we wish to protect?" There is also considerable inertia when attempts are made to move regulatory approaches forward in the light of new technical advances, and a lack of appropriate management systems for auditing past decisions in the light of new knowledge.

10.3 Ecotoxicity test methodology

10.3.1 Do current test methods provide useful information?

Current test methods and frameworks do provide useful information, but it is limited and needs to be recognized as such. For example, only the 72-h algal growth test[23] currently provides information at a population level of biological organization in the form of the intrinsic rate of population increase. However, the quality of even this information could be improved if effects on survival were separated from effects on growth by either counting dead cells or re-inoculating algae into undosed media.

In all ecotoxicity tests, it was agreed that one value is generally insufficient to summarize the results of a statistical analysis. For example, reporting the EC50 *and* the slope of the dose–response curve provides considerably more information than reporting just the EC50.[12] Use of the NOEC was considered to be an unacceptable waste of available information from any test, although the estimation of "benchmark concentrations"[3] can be of value in risk assessment.

There was also agreement that, in certain areas, such as effluent ecotoxicity testing in Europe, further test systems would need to be used to help understand the long-term sublethal effects of toxic effluents.

10.3.2 Could more useful information be obtained from current test methods?

Certainly, more information can be obtained from current tests, particularly since the availability of computing power has increased enormously in recent years. Access to good quality software and statisticians who can operate it may be a problem, but the most important question is whether environmental regulators would actually use additional information. The answer to this will be specific to individual regulators and the organizations and legal frameworks within which they work.

As a minimum, LC/EC50 values plus their 95% fiducial limits should be reported for each observation period (e.g., 24-, 48, – 72- and 96-h) so that a toxicity or time to event curve can be constructed.[24] Information that might also be useful to a regulator includes an estimate of the true no effect concentration (NEC) rather than a NOEC, some idea of whether continuation of an ecotoxicity test beyond its usual duration might produce a more sensitive result, and the collection of data in a way that allows potential effects on population growth to be determined. For example, data from water flea reproduction tests are currently reported as median lethal concentrations and mean number of offspring produced per surviving female.[25] If the same data were reported as age-specific survival and fecundity, standard life table analyses could be used to estimate the intrinsic rate of population increase.[26]

Environmental regulators should also be sent all raw data, so that results can be reanalyzed should new techniques emerge. Ideally, these data would be made available to all via the Internet, although it is recognized that the ownership and use of raw data is a controversial subject. A more stringent approach to acceptable control variability would also be useful in discriminating among tests that are conducted well or poorly. Reporting the residual coefficient of variation of the control would provide this information.

10.3.3 Would alternative test designs provide more useful information?

"Better" tests that provide more focused information or take account of the properties of the substance being tested can always be designed. However, because we do not currently use all of the information provided by current tests, it might be best to begin by using existing information, rather than designing new tests.

10.4 Chemical risk assessment

10.4.1 Is information from ecotoxicity tests used optimally by chemical risk assessors?

Current risk assessment practice for substances such as pesticides and industrial chemicals appears to protect the environment adequately because it is conservative — not because it is scientifically based. Certainly, it does not use probabilistic risk-based techniques, although it is couched in these terms. Problems of interpretation after the use of safety or extrapolation factors usually arise in the "gray zone" for chemicals that are neither clearly safe nor clearly and unacceptably hazardous. Risk could be addressed more explicitly by estimating the uncertainty around estimates of both environmental effect and environmental concentration (e.g., by "estimation of extrapolation error"[27]), or by bootstrapped estimates of species sensitivity distributions.[28]

Many questions also exist about both the "risk" and "assessment" components of a risk assessment. Many researchers question the criteria used for selecting measurement and risk assessment endpoints,[29] and the perception of risk often appears to vary among the public, regulators and industrialists.[30] Current summary statistics probably do not help this situation, as their meaning is unclear to many. It is easy for the public to confuse an NOEC with an NEC and it is difficult for them to understand what a 96-h LC50 is and how dividing it by a multiple of 10 helps to protect the environment. The expression of ecotoxicity data as changes in risk through time and at sensitive life history stages, changes in life expectancy after exposure to contaminants, effects on population recovery time or on the sustainability of population production, may be a better way of communicating this information to a non-technical audience.

Effluent toxicity testing and retrospective toxicity assessments of historical contamination can present problems because the maximum concentration that can be tested is limited to 100% of the undiluted sample. This might mean that such tests are inherently less conservative than ecotoxicity tests with pesticides and industrial chemicals in which concentrations far above likely environmental concentrations can be tested. Another problem is that the chemical, or chemicals, causing observed biological effects in bioassays of environmental media often remains unknown. Risk assessors in these fields would benefit from any additional information that allowed them to distinguish the toxic effects of different samples. Different *rates* of toxicity through time are one way in which such samples can be distinguished.

10.5   Time to event methods

10.5.1   How can time to event approaches be used to improve the interpretation of ecotoxicity tests?

The results from ecotoxicity tests are used to:

- Help make commercial and regulatory management decisions about whether a substance should be manufactured and marketed, and how its environmental release should be controlled.
- Communicate hazard and risk to nontechnical managers and the public.

Suggestions for new approaches should be judged against these objectives to determine whether they are cost-effective. Several different advantages of time to event analyses of ecotoxicity data that meet at least one of the above objectives were suggested:

- Biological effects can be estimated over any time period using TTE analysis, not just at the end of an arbitrarily fixed exposure period.[20]
- Currently, acute-to-chronic ratios are often based on very different acute and chronic measurement endpoints (e.g., survival versus reproduction). The extrapolation of short-term to long-term effects (acute to chronic extrapolation) is placed on a firmer scientific foundation if trends in toxic effect through time in short-term tests are used to predict potential long-term effects on similar endpoints. Although there are dangers in extrapolating beyond the range of available data, this approach would be a sensible way of prioritizing chemicals, effluents or other environmental media that should be assessed in longer-term tests.[18]
- The likelihood that an important life history event, such as spawning or hatching, will be delayed or promoted can be estimated. Associated with this, the reduction in life expectancy of valued ecosystem components exposed to contaminants can be expressed with an estimate of uncertainty.[31]
- Outputs from time to event analyses can be used as inputs for fish stock models or other types of population models.[20] These can then be used to predict the intrinsic rate of population increase and determine when this is >0 (population increasing) or <0 (population declining to extinction). The time to population or system recovery can also be estimated by TTE analysis, as can appropriate remediation levels. However, it is recognized that the intrinsic rate of increase may vary between laboratory and field, due to a variety of feedback

mechanisms in the field that are not present in the laboratory. The use of time to event analyses does not remove the need for careful field monitoring of laboratory predictions.

- Time-varying exposure can be analyzed by TTE approaches. This may occur if releases to the environment vary (e.g., the toxicity of releases from a sewage treatment works), or because a mobile organism is exposed only intermittently, such as an insectivorous bird feeding upon prey in pesticide-contaminated and -uncontaminated fields.[32] Knowledge of modes of toxic action would be a useful complement for such analyses, allowing an investigator to distinguish between those chemicals that are likely to produce rapid irreversible effects and those likely to produce fully reversible effects.
- The results from a TTE analysis can be expressed as a three-dimensional concentration-time-response surface, which is a useful visual tool for communicating risk.[27]
- Results from TTE analysis can also be expressed as changes in relative risks through time or as a lifetime increase or decrease in risk if a proportional hazards model is used.[21]
- The effect on toxicity of important covariates such as sex, organism weight and genotype can be examined efficiently.[33]
- Finally, improved precision and enhanced power can be obtained from a time to event analysis, even if a fixed time LC/EC value remains the chosen summary statistic (see Chapter 5). However, this will be of practical use only if LC50 values with narrow confidence intervals are treated by regulators in a different way from those with wide confidence intervals.

In retrospective impact assessments, the use of population modeling is more widespread than in predictive assessments, because protecting biodiversity at specific locations is more explicitly recognized as a risk assessment objective. It therefore makes sense to analyze environmental bioassay results in a way that produces information that can be fed into these models. The temporal dimension of toxicant exposure is also often much clearer in retrospective risk assessments, because an assessor may have some direct measurements of changes in contaminant concentration and biological effects through time. Again, this makes the use of time to event analysis a sensible choice.

Members of the pesticides working group were less enthusiastic about the use of time to event analysis, especially for acute data. Their reservations were that:

1. Although time to mortality is recorded in acute pesticide tests at 24, 48, 72 and 96 h, and can be used to plan further tests, pesticide regulators are not interested in anything other than the 96-h LC50 for comparison with a Predicted Environmental Concentration, be-

cause the first tier in pesticide risk assessment is simply designed to act as a possible trigger for further tests.

2. Delayed effects that occur after the end of exposure to pesticides add an additional layer of uncertainty to temporal effects;

3. The use of large safety factors to account for uncertainties such as interspecies differences in sensitivity will swamp any gains made by use of time to event analysis. The pesticide working group did see more potential for the use of time to event analysis in longer-term tests such as the *Daphnia* 21-d reproduction test, and in specifically designed higher tier tests.

In contrast, in the U.S., members of the ECOFRAM pesticide risk assessment group are keener on the use of TTE analysis for acute rather than chronic studies (MC Newman, pers. comm.). This is because it allows the analysis of effects caused by one or more brief, pulsed exposures that might occur for a few hours after a storm event. However, the experience of members of the Wellesbourne workshop pesticides working group was that pesticide regulators asked for pulsed experiments to be performed, rather than time to event analyses, because they wished to confirm that delayed effects from an initial exposure were not entirely responsible for the total level of observed effects. This is the "latency" problem: although organisms may survive a brief pulse, they could die later because of irrecoverable damage. It was pointed out that the latency problem quite clearly also exists in fixed time ecotoxicity test analyses (e.g., a 96-h LC50), and that the inclusion of time delays in the analysis of toxicity data probably first requires the adoption of a time to event approach.

Some delegates also disagreed with the pesticide group over the issue of uncertainties in other areas of risk assessment. It was felt that we should not ignore possible improvements in one area of science because of large uncertainties elsewhere. Progress in science is usually incremental, and many small improvements in risk assessment approaches should lead to considerable progress over time. It was also thought that environmental regulators who are not currently using available data in the optimal way should be educated to do so.

10.5.2 Should current test designs be modified to optimize time to event analyses?

The optimal length of ecotoxicity tests and the optimal number of observations on each experimental unit through time will depend in part on our knowledge of both the chemical and biological characteristics of the test system. Physico-chemical considerations, such as the partition coefficient in relation to the size of the test organism, will determine how rapidly a contaminant is absorbed, and may suggest the extension of a test if organisms are large (e.g., fish). The weight of animals such as fish and water fleas could be recorded at the end of experiments so that mechanistic models that incorporate time[34] can be used most effectively. A cost–benefit analysis would be useful to compare the additional costs of taking these measurements with the benefits of increased information for decision making.

Observations should be spaced to occur at periods of maximum change in effect, so prior knowledge of the likely timing of effects is very useful. As a rule of thumb, one or two observations per day, excluding weekends, would probably be sufficient for most standard long-term ecotoxicity tests lasting for more than 10 working days. Measurements in "log time" (e.g., 0, 1.5, 3, 6, 12, 24 and 48 h) may be most efficient for some tests. If effects continue to occur at the end of a test, it should be continued, if possible, until an asymptote is reached. However, experimenters may then encounter difficulties due to the interaction of toxicity and starvation in acute tests. With time to event analysis, more information will usually be gained if replication at each concentration is converted into more concentrations with fewer, or single, replicates. However, the potential effects of pseudoreplication[35] would need to be assessed for each test.

Data on mortality and reproduction should never be discarded or composited if a time to event analysis is to be performed. As stated previously, the *Daphnia* 21-d reproduction test[25] produces data that could be used in population models, but the way that these data are reported usually prevents this useful exercise. Another water flea test, with *Ceriodaphnia*, suffers from a different problem. In this 7-day test, reproduction usually does not begin until day 4, thus, rather few data points for reproduction are available at the end of the test.[36] An extension of the test duration to 10 days could solve this problem.

Since environmental regulators are usually interested in only small levels of effect, estimation of an EC50 appears wasteful of replicates. If only "small" effects are acceptable, most experimental units should be exposed to low concentrations. This design would not be appropriate if a risk assessor was interested in the concentration–time–response relationships that occur after an accidental spillage of high chemical concentration. As usual, specific focused experimental designs can be prepared only for specific focused scientific or regulatory questions.

Finally, experimenters may wish to identify multiple relevant endpoints within their tests that are amenable to time to event analysis. For example, in the Fish Early Life Stage test,[37] time to death, time to hatch and time to swim up can all be measured in the same experiment.

10.5.3   Should new tests be designed to allow time to event analyses?

There is no apparent need for additional test guidelines for standard ecotoxicity tests. Nearly all current guidelines produce data that can be analyzed by time to event analysis. When specific tests are designed to answer specific questions, the dimension of time should be considered.

10.6 Conclusions and recommendations

1. Environmental legislation should remain general and flexible and should not specify the use of analytical techniques, such as the NO-EC, that may become obsolete. Legislation should instead encourage the use of current best practice.
2. Time is an undervalued dimension in current statistical analyses of ecotoxicity tests, but one that chemical risk assessors often need to consider. Chemical risk assessors must define their objectives clearly and be more explicit in defining those objectives that require the analysis of time-dependent toxicity.
3. Time to event analyses are commonly used in other disciplines, such as medicine and engineering, to analyze data similar to those obtained from ecotoxicity tests.
4. Because all data points are used in the analysis, time to event analyses are usually more powerful than the statistical techniques currently used for hypothesis testing and EC estimation.
5. Alternatively, fewer organisms can be used in ecotoxicity tests to achieve a level of power similar to that currently attained by standard methods.
6. Time to event analysis provides a more rational basis for acute to chronic extrapolations, and for prioritizing chronic studies.
7. Time to event analysis allows the examination of changes in toxic hazard over time, or during different life stages.
8. Time-varying exposure can be analyzed more effectively by using time to event analysis.
9. Time to system recovery can be predicted in retrospective impact assessments.
10. Biological population parameters, such as the intrinsic rate of population increase, can be estimated from time to event analyses.
11. Toxic risks can be communicated to the public more effectively by using time to event analyses to estimate summaries such as the relative risk of exposure, or reduced life expectancy after exposure to toxicants.
12. Analyses of standard data sets using time to event approaches are required to validate these methods and as an educational tool for all those involved in the chemical risk assessment process. These analyses should be used to identify the advantages and disadvantages of time to event analysis in risk assessment.

Acknowledgements

We thank all of the Wellesbourne workshop participants, listed in the Appendix, for contributing their time and expertise. We also thank two anonymous referees whose comments improved the clarity of this chapter.

Appendix: Workshop Participants

Participant	Organization
Ian Barber	AgrEvo U.K. Ltd., U.K.
Jacques J.M. Bedaux	Vrije Universiteit, The Netherlands
Ruth Boumphrey	The Environment Agency, U.K.
Nick Cartwright	WRc plc, U.K.
Peter Chapman	Zeneca Agrochemicals, U.K.
Mark Crane	Royal Holloway University of London, U.K.
Peter Delorme	Pest Management Regulatory Agency, Canada
Philip Dixon	Iowa State University, USA
Peter Dohmen	BASF AG, Germany
John Fenlon	Horticulture Research International, U.K.
Dave Forrow	The Environment Agency, U.K.
Ann Gould	Statistical Consultant, U.K.
Mick Hamer	Zeneca Agrochemicals, U.K.
Claus Hansen	Environmental Protection Agency, Denmark
Chris Harbron	Unilever Research, U.K.
Chris Karman	TNO Institute of Environmental Sciences, The Netherlands
Herbert Koepp	BBA, Germany
Bas Kooijman	Vrije Universiteit, The Netherlands
Bryan Manly	Western EcoSystem Technology Inc., USA
Eddie McIndoe	Zeneca Agrochemicals, U.K.
Reinhard Meister	TFH Berlin, Germany
Michael C. Newman	Virginia Institute of Marine Science, USA
Alan Sharpe	Brixham Environmental Laboratory, U.K.
Tim Sparks	Institute of Terrestrial Ecology, U.K.
Eric Vindimian	INERIS, France
Paul Whitehouse	WRc plc, U.K.
Douglas Wilson	ADAS, U.K.

References

1. Skalski, J.R., Statistical inconsistencies in the use of no-observed-effect-levels in toxicity testing, in, *Aquatic Toxicology and Hazard Assessment: Fourth Conference*, Branson, D.R. and Dickson, K.L., Eds., Philadelphia, PA, 1981, 377.
2. Bruce, R.D. and Versteeg, D.J., A statistical procedure for modeling continuous toxicity data, *Environ. Toxicol. & Chem.*, 11, 1485, 1992.
3. Crump, K.S., A new method for determining allowable daily intakes, *Fund. & Appl. Toxicol.*, 4, 854, 1984.
4. Hoekstra, J.A. and van Ewijk, P.H., Alternatives for the no-observed effect level, *Environ. Toxicol. & Chem.*, 12, 187, 1993.

5. Kooijman, S.A.L.M., Parametric analyses of mortality rates in bioassays, *Wat. Res.*, 15, 107, 1981.
6. Kooijman, S.A.L.M., An alternative for NOEC exists, but the standard model has to be abandoned first, *Oikos*, 75, 310, 1996.
7. Laskowski, R., Some good reasons to ban the use of NOEC, LOEC and related concepts in ecotoxicology, *Oikos*, 73, 140, 1995.
8. Noppert, F., Leopold, A. and van der Hoeven, N., *How to Measure No Effect: Towards a New Measure of Chronic Toxicity in Ecotoxicology*, BKH Consulting Engineers, Delft, The Netherlands, 1994.
9. Stephan, C.E. and Rogers, J.W., Advantages of using regression to calculate results of chronic toxicity tests, in, *Aquatic Toxicology and Hazard Assessment: Eighth Symposium*, Bahner, R.C. and Hansen, D.J., Eds., Philadelphia, PA, 1985, 328.
10. Grandy, N.J., Role of the OECD in chemical control and international harmonization of testing methods, in, *Fundamentals of Aquatic Toxicology*, Second edition, Rand, G.M., Ed., Taylor & Francis, Washington, D.C., 1995, 763.
11. Pack, S., *A Review of Statistical Data Analysis and Experimental Design in OECD Aquatic Toxicology Test Guidelines*, Organisation for Economic Cooperation and Development, Paris, France, 1993.
12. Chapman, P.F., Crane, M., Wiles, J., Noppert, F. and McIndoe, E., Improving the quality of statistics in regulatory ecotoxicity tests, *Ecotoxicology*, 5, 169, 1996.
13. OECD, *Report of the OECD Workshop on Statistical Analysis of Aquatic Toxicity Data*, Braunschweig, Germany, 15-17 October 1996, Organisation for Economic Co-operation and Development, Paris, France, 1997.
14. Dixon, P.M. and Newman, M.C., Analysing toxicity data using statistical models of time-to-death: an introduction, in, *Metal Ecotoxicology: Concepts and Applications*, Newman, M.C. and McIntosh, A.W., Eds., Lewis, Chelsea, MI, 1991, 207.
15. Newman, M.C. and Aplin, M., Enhancing toxicity data interpretation and prediction of ecological risk with survival time modeling: an illustration using sodium chloride toxicity to mosquitofish (*Gambusia holbrooki*), *Aquat. Toxicol.*, 23, 85, 1992.
16. Newman, M.C., Diamond, S.A., Mulvey, M. and Dixon, P., Allozyme genotype and time to death of mosquitofish, *Gambusia affinis* (Baird and Girard) during acute toxicant exposure: a comparison of arsenate and inorganic mercury, *Aquat. Toxicol.*, 15, 141, 1989.
17. Newman, M.C., Keklak, M.M., Doggett, M.S., Quantifying animal size effects on toxicity: a general approach, *Aquat. Toxicol.*, 28, 1, 1994.
18. Sun, K., Krause, G.F., Mayer, F.L. Jr., Ellersieck, M.R. and Basu, A.P., Predicting chronic lethality of chemicals to fishes from acute toxicity test data: theory of accelerated life testing, *Environ. Toxicol. & Chem.*, 14, 1745, 1995.
19. Muenchow, G., Ecological use of failure time analysis, *Ecology*, 67, 246, 1986.
20. Newman, M.C., *Quantitative Methods in Aquatic Ecotoxicology*, Lewis, Boca Raton, FL, 1995.
21. Cox, D.R. and Oakes, D., *Analysis of Survival Data*, Chapman and Hall, London, 1984.
22. Lee, E.T., *Statistical Methods for Survival Data Analysis*, Lifetime Learning Publications, Belmont, CA, 1980.

23. OECD, *Guidelines for Testing of Chemicals, Section 2: Effects on Biotic Systems. 201: Alga, Growth Inhibition Test,* Organisation for Economic Co-operation and Development, Paris, France, 1984.

24. Sprague, J.B., Measurement of pollutant toxicity to fish. I. Bioassay methods or acute toxicity, *Water Research,* 3, 793, 1969.

25. OECD, *Guidelines for Testing of Chemicals, Section 2: Effects on Biotic Systems, 202: Daphnia spp., Acute Immobilisation Test and Reproduction Test,* Organisation for Economic Co-operation and Development, Paris, France, 1984.

26. Walthall, W.K. and Stark, J.D., A comparison of acute mortality and population growth rate as endpoints of toxicological effect, *Ecotoxicol. & Environ.Safety,* 37, 45, 1997.

27. Suter, G.W., II., *Ecological Risk Assessment,* Lewis, Chelsea, MI, 1993.

28. Jagoe, R.H. and Newman, M.C., Bootstrap estimation of community NOEC values, *Ecotoxicology,* 6, 293, 1997.

29. Cairns, J., Jr., The myth of the most sensitive species, *Bioscience,* 36, 670, 1986.

30. Douben, P.E.T., *Pollution Risk Assessment and Management: A Structured Approach,* John Wiley and Sons, Chichester, UK, 1998.

31. Leslie, P.H., Tener, J.S., Vizoso, M. and Chitty, H., The longevity and fertility of the Orkney vole, *Microtus orcadensis,* as observed in the laboratory, *Proc. Zoological Soc. London,* 125, 115, 1955.

32. Widianarko, B. and van Straalen, N., Toxicokinetics-based survival analysis in bioassays using nonpersistent chemicals, *Environ. Toxicol. & Chem.,* 15, 402, 1996.

33. Diamond, S.A., Newman, M.C., Mulvey, M., Dixon, P.M. and Martinson, D., Allozyme genotype and time to death of mosquitofish, *Gambusia affinis* (Baird and Girard), during acute exposure to inorganic mercury, *Environ. Toxicol. & Chem.,* 8, 613, 1989.

34. Kooijman, S.A.L.M. and Bedaux, J.J.M., *The Analysis of Aquatic Toxicity Data,* VU University Press, Amsterdam, the Netherlands, 1996.

35. Hurlbert, S.H., Pseudoreplication and the design of ecological field experiments, *Ecol. Mono.,* 54, 187, 1984.

36. USEPA, *Methods for Estimating the Chronic Toxicity of Effluents and Receiving Waters to Freshwater Organisms,* EPA-600/4-89-001, United States Environmental Protection Agency, Environmental Monitoring Systems Laboratory, Cincinnati, OH, 1989.

37. OECD, *Guidelines for Testing of Chemicals, Section 2: Effects on Biotic Systems. 210: Fish, Early-Life Stage Toxicity Test,* Organisation for Economic Co-operation and Development, Paris, France, 1992.

chapter 11

Conclusions

Mark Crane, Peter F. Chapman and Michael C. Newman

The chapters in this book show how time to event methods have been used in many different disciplines, and their clear potential for use in the risk assessment of chemicals and contaminated media. Statistical approaches currently used to analyze ecotoxicity data grew out of the "end of pipe" regulatory context of the 1960s and 1970s, when control of acutely lethal point source discharges was a priority.[1] Now there are higher demands on data interpretation, as environmental regulation grows toward a risk assessment context. More powerful statistical methods are required to meet these higher demands, and the capability to perform these analyses has grown dramatically with increases in computing power and software availability. More and more, risk assessors wish to decompose uncertainty in their assessments into areas of reducible versus irreducible uncertainty, and to minimize the use of arbitrary safety factors. More powerful time to event approaches that take account of the time course of toxicity can reduce uncertainty in an important area for risk assessors.

The availability of literature to describe different time to event approaches has also increased, and is widely used in other fields. Cox proportional models, among the most commonly used semiparametric approaches in time to event analysis, are based on an original paper[2] that is one of the top ten cited articles in science today. Fully parametric and nonparametric approaches are also available,[3] as described in Chapter 1. Although a reasonable level of numeracy is required to use some of the more sophisticated time to event approaches, Chapter 2 shows that even the modestly numerate can gain some rewards from use of time to event in standard software packages.

The use of time to event approaches offers some clear advantages. Time to event methods explicitly address the twin causes of toxic effects — the intensity and duration of exposure to hazardous chemicals, and better use is made of the data gathered, often at great expense, from toxicity experiments. Diverse end points in time can be addressed with time to event approaches, and individual organism characteristics can be incorporated as covariables, as shown in Chapter 3.

Because more data are generated, it is possible to select more appropriate statistical models, and to relate these models to population level effects[4] and epidemiological and exposure models, as shown in Chapters 3 and 6. Data from field studies can be analyzed as readily as data from laboratory studies, as shown in Chapter 8. Because of this flexibility, both prospective and retrospective risk assessments can benefit from use of time to event analyses. Time to event approaches are also likely to be most effective for analysis of toxicity from pulsed exposures, and of latent toxic effects emerging after exposure has ceased, because both of these phenomena are time-related. In addition, projection of likely toxic effects in short-term experiments to effects over the long-term is possible, as shown in Chapter 4.

Expression of results from toxicity experiments as *risks* or as reduced life expectancy, rather than as LC50 or NOEC summaries, is likely to aid in communication of the absolute or relative dangers of particular chemical exposure scenarios, as described in Chapters 2 and 3. Finally, from an animal welfare point of view, fewer animals may be necessary in some experimental designs if the increased statistical power from time to event approaches described in Chapter 5 is traded off against reduced numbers of test organisms.

Time to event approaches do have some negative points that count against them. Currently, they are not the approach taken by most environmental toxicologists, and recording of data at several different time points is more time-consuming than recording at one final time point. However, interval censored data can be analyzed by several time to event approaches, including the life tables used for many years by actuaries and ecologists.[5-10] In addition, video technology is now inexpensive and accessible and could make the recording of exact results from toxicity tests much easier.[11] Pseudoreplication,[12] already a factor in most toxicity test designs, may be more important when gathering more data in a time to event design. We are not sure how great this problem would be in practice, so this is an area where research is required.

Further research is also required into the use of time to event approaches for analyzing pulsed exposures and latent effects, and optimized laboratory experimental designs for investigating the toxic effects of exposure to low concentrations of chemicals. Development of user-friendly software tailored toward the time to event needs of environmental risk assessors would also be useful.

The time course of toxicity has not always been the poor relation of exposure intensity in environmental toxicology. In the 1960s, John Sprague wrote of its importance,[13] and thresholds for median lethal *times* (LT50) were frequently reported. Since then, more powerful approaches than the LT50 have been developed and used in medicine, agriculture, engineering and ecology, as described in Chapters 7, 8, 9 and 10. There is no doubt that risk assessment in the field of environmental toxicology and chemistry can also be improved by use of time to event models.

References

1. Newman, M.C., *Quantitative Methods in Aquatic Ecotoxicology*, Lewis, Boca Raton, FL, 1995.
2. Cox, D.R., Regression models and life tables (with discussion), *J. Royal Stat. Soc.*, B34, 187, 1972.
3. Cox, D.R. and Oakes, D., *Analysis of Survival Data*, Chapman & Hall, London, 1984.
4. Caswell, H. and John, A.M., From the individual to the population in demographic models, in *Individual-based Models and Approaches in Ecology*, DeAngelis, D.L. and Gross, L.J., Eds., Chapman & Hall, New York, NY, 1992, 36.
5. Pyke, D.A. and Thompson, J.N., Statistical analysis of survival and removal experiments, *Ecology*, 67, 240, 1986.
6. Leslie, P.H., Tener, J.S., Vizoso, M. and Chitty, H., The longevity and fertility of the Orkney vole, Microtus orcadensis, as observed in the laboratory, *Proc. Zool. Soc. London*, 125, 115, 1955.
7. Daniels, R.E. and Allan, J.D., Life table evaluation of chronic exposure to a pesticide. *Can. J. Fish. & Aquat. Sci.*, 38, 485, 1981.
8. Day, K. and Kaushik, N.K., An assessment of the chronic toxicity of the synthetic pyrethroid, fenvalerate, to *Daphnia galeata mendotae*, using life tables, *Environ. Poll.*, 44, 13, 1987.
9. Pesch, C.E., Munns, W.R. and Gutjahr-Gobell, R., Effects of a contaminated sediment on life history traits and population growth rate of Neanthes arenaceodentata (*Polychaeta: Nereidae*) in the laboratory, *Environ. Toxicol. & Chem.*, 10, 805, 1991.
10. Martínez-Jerónimo, F., Villaseñor, R., Espinosa, F. and Rios, G., Use of life-tables and application factors for evaluating chronic toxicity of Kraft mill wastes on *Daphnia magna*, *Bull. Environ. Contam. & Toxicol.*, 50, 377, 1993.
11. Baartrup, E., Bayley, M., Sørensen, F.F. and Toft, G., Can animal behaviour predict population level effects?, in *Forecasting the Environmental Fate and Effects of Chemicals*, Rainbow, P.S., Hopkin, S.P. and Crane, M., Eds., Wiley, Chichester, U.K., 2001, 139.
12. Hurlbert, S., Pseudoreplication and the design of ecological field experiments, *Ecol. Mono.*, 54, 187, 1984.
13. Sprague, J.B., Measurement of pollutant toxicity to fish. I. Bioassay methods for acute toxicity, *Water Res.*, 3, 793, 1969.

Index

Milton Keynes UK
Ingram Content Group UK Ltd.
UKHW040055071024
449327UK00019B/567